Digital Monuments in Digital Landscape

Daniel Bešina

Digital Monuments in Digital Landscape

PETER LANG
Berlin · Bruxelles · Chennai · Lausanne · New York · Oxford

Library of Congress Cataloging-in-Publication Data
Names: Bešina, Daniel, author.
Title: Digital monuments in digital landscape / Daniel Bešina.
Description: Berlin ; New York : Peter Lang, [2025] | Includes bibliographical references and index.
Identifiers: LCCN 2025015829 (print) | LCCN 2025015830 (ebook) | ISBN 9783631936221 hardback | ISBN 9783631936238 ebook | ISBN 9783631936245 epub
Subjects: LCSH: Monuments--Conservation and restoration | Cultural landscapes | Historic preservation--Technological innovations | Historic sites--Reconstruction
Classification: LCC CC135 .B476 2025 (print) | LCC CC135 (ebook) | DDC 363.6/9--dc23/eng/20250408
LC record available at https://lccn.loc.gov/2025015829
LC ebook record available at https://lccn.loc.gov/2025015830

Bibliographic Information published by the Deutsche Nationalbibliothek

The Deutsche Nationalbibliothek lists this publication in the Deutsche Nationalbibliografie; detailed bibliographic data is available on the internet at http://dnb.d-nb.de.

This publication was created with the support of the Scientific Grant Agency of the Ministry of Education, Science, Research, and Sport of the Slovak Republic under the VEGA project 1/0149 and Constantine the Philosopher University in Nitra

ISBN 978-3-631-93622-1 (Print)
ISBN 978-3-631-93623-8 (E-PDF)
ISBN 978-3-631-93624-5 (E-PUB)
DOI 10.3726/b22800

Published by Peter Lang GmbH, Berlin (Germany)

info@peterlang.com

This publication has been peer reviewed.

www.peterlang.com

This book is dedicated to my parents and grandparents for their lifelong support and guidance.

Author: PhDr. Daniel Bešina, PhD.

Reviewers:
doc. et doc. PaedDr. Pavel Hrončék, PhD.,
Mgr. Martin Tomášek, Ph.D.

Language proofreading and translation into English:
PhDr. Ľudmila Vanková

Extent:

Published by:
This publication was created with the support of the Scientific Grant Agency of the Ministry of Education, Science, Research and Sport of the Slovak Republic under the VEGA project 1/0149

Contents

List of Figures, Diagrams and Tables

Figures

Diagrams

Tables

Acknowledgment

I would like to express my sincere gratitude to everyone who significantly contributed to the creation of this monograph and provided me with professional assistance, support, and inspiration during its preparation.

First and foremost, I extend my thanks to the Scientific Grant Agency of the Ministry of Education and Constantine the Philosopher University in Nitra for their financial and material support, which made this work possible.

My heartfelt thanks go to the members of the scientific team – Stanislava Bönde Gogová, Daniel Miček, and Radomír Sabol – for their valuable collaboration, insightful discussions, and professional support throughout the entire process.

Special thanks are owed to Pavol Hronček for his expert consultations, which significantly contributed to the accuracy and quality of this work.

Finally, I extend my gratitude to my family and friends for their support, patience, and encouragement during the work on this monograph.

This monograph would not have been possible without the cooperation and efforts of many individuals, to whom I am deeply grateful.

List of Abbreviations

3D	Three-Dimensional
AHP	Analytic Hierarchy Process
ALS	Airborne Laser Scanning
AR	Augmented Reality
BIM	Building Information Modeling
CAD	Computer-Aided Design
CH	Cultural Heritage
DTM	Digital Terrain Model
DMR	Digital Model of Relief
EHL	European Heritage Label
ETS	European Treaty Series
EU	European Union
GIS	Geographic Information Systems
GNSS	Global Navigation Satellite System
HBIM	Historic Building Information Modeling
HLIM	Heritage Landscape Information Modeling
HMM	Historic Monument Management
ICOMOS	International Council on Monuments and Sites
IFC	Industry Foundation Classes
IT	Information Technology
LiDAR	Light Detection and Ranging
LIM	Landscape Information Modeling
LOD	Level of Detail

M&E	Mechanical and Electrical systems
RPAS	Remotely Piloted Aircraft Systems
RGB	Red, Green, Blue (color model)
SfM	Structure from Motion
TLS	Terrestrial Laser Scanning
UAS	Unmanned Aerial Systems
UAV	Unmanned Aerial Vehicle
UNESCO	United Nations Educational, Scientific and Cultural Organization
UV	Unwrapping Vertices (used in 3D modeling)
VR	Virtual Reality
VRCO	Virtual Reality Culture Organization
WHC	World Heritage Committee

Introduction

The landscape, intertwined with a web of natural beauty and human works, is a living testimony of millennia of development, change and persistence. The monuments situated in it are like precious gems, hidden in the ground or on the surface, waiting to be discovered and appreciated. Every stone, every wall and every pillar that has survived this immense time tells a story about the people who lived here, about their efforts, successes and downfalls. Exploring these monuments is like uncovering ancient manuscripts that carry invaluable knowledge about the past that can illuminate our present and guide the future.

The cultural landscape in which these monuments are located is not just a passive background to their stories, but is a partaker, who actively shapes their meaning. It is scenery where great historical events took place, where cultures were born, where the destinies of nations intersected. Every piece of land, every valley and hill bears traces of the past that have shaped not only the physical nature of the country but its spiritual dimension as well. Therefore, exploring the monuments in the landscape is extremely important, as it allows us to understand how the environment affected people and how people changed their environment, creating the cultural landscape that we try to comprehend today.

The presented publication "Digital Monuments in Digital Landscape" focuses on this relationship between monuments and the landscape. It explores how modern digital tools and technologies can contribute to conservation and rebirth of this heritage. If monuments in the landscape represent our collective memories and identity, then digital technologies are like a mirror that allows us to see those memories again, to relive them and preserve for future generations. This monograph provides not only an insight into how monuments can be captured and presented through digital tools, but also how this technology can be used to protect and reconstruct them. At a time when many monuments are facing the risk of extinction, either as a result of natural degradation or due to human activity, new ways to conserve them are essential to be quested. Digital technologies can offer the opportunity not only to document the current state of these monuments, but also to virtually reconstruct them and make them available to the public in the form they had centuries ago.

Exploring the landmarks in the countryside and integrating them into digital world is like finding a new path between the past and the future. This connection allows us both to maintain contact with our past and to transfer these precious values to new generations, who will face new challenges, but will also have a firm support in the knowledge and heritage of their predecessors.

The monograph is aimed in acquainting the professional public with current trends and challenges in the field of monument care and at the same time in bringing these topics closer to the wider lay public. The monograph provides a deep insight into how the landscape monuments are significant as cultural and historical components shaping the identity and aesthetic character of the regions. Nowadays, when technology plays a key role in many areas of life, to explore how it can contribute to the sustainability and protection of our cultural heritage is essential.

Introductory chapters are focused on the theoretical foundations, where the importance of monuments in the landscape context, their role in creating of cultural and historical identity and their continuing impact on the aesthetic character of regions is discussed in detail. The following part explores the landscape as a monument, emphasizing how the long-term interaction of natural and human factors has shaped the historical landscapes that are recently perceived as a relevant part of the cultural heritage.

In next chapters practical aspects of digital monument care are discussed in detail. Modern methods of digital recording, documenting and analyzing the monuments are being developed here, including the use of 3D scanning, drones and other advanced technologies, which enable accurate and efficient documentation of monuments. The monograph deals with the process of digital modeling and reconstruction of historic objects (BIM) in detail, thus providing guidance to revive and conserve objects that are no longer physically present or are seriously damaged. The chapters on virtual reality and its use in the monument care demonstrate how these technologies allow the public to experience and learn about monuments in their original form, opening up new possibilities for education, cultural tourism and interactive presentation of historical heritage.

The monograph further discusses the importance of digital heritage for future generations, analyzing how digital technologies can contribute to the conservation and accessibility of monuments at the time when many of them are threatened with extinction due to natural or human influences. The discussion of the legal and ethical aspects of digital monument care is important as well, where the issues of copyright, personal data protection and ensuring access to digital archives are addressed. The next chapters emphasize the need for an interdisciplinary approach that brings together experts from various fields of science to create complex and sustainable solutions in the cultural heritage digital protection. The chapter on education and extending of awareness about digital monument care shows how important it is to involve the public in heritage conservation processes via modern educational programs and interactive platforms.

The monograph is contributing to the above mentioned issue by its own design of virtual reconstruction evaluation system for monuments and historical landscapes. The uncertainty classification system for 3D reconstructions is designed to provide a structured framework for evaluating the various aspects that affect the accuracy and authenticity of these digital models. The classification system includes several criteria, such as changes in the geo-relief, preservation of anthropogenic structures, historical mapping or availability of 3D digital documentation. Using this system, experts can better identify and quantify reconstruction uncertainties, thus increasing confidence in final models and visualizations. This approach is extremely useful not only for research and conservation, but also for transparent communication with the public, which allows a better understanding of the value and authenticity of the monuments presented.

The publication also contains specific case studies that can demonstrate the ways how theoretical knowledge can be applied in the practice of the monument reconstruction and protection, while addressing the issue of uncertainty and accuracy in digital reconstruction. In the final chapters, the authors discuss the future of digital monument care, identify the main trends that will shape the field, and discuss the challenges associated with digital archives sustainability, growing need for interdisciplinary collaboration, and constantly changing technological conditions.

Developing digital technologies can offer new possibilities for the monument protection and presentation. Virtual reconstructions make it possible not only to preserve the appearance and atmosphere of historical places for future generations, but also to make them accessible to the wider public. The reconstruction in virtual space provides a unique opportunity to explore monuments that are physically inaccessible or damaged. It allows creating of dynamic and interactive models that can serve as tools for education, research and tourism. Virtual reality thus represents a new dimension in the monument care, which can interlink the past with the future and proffer new forms of public engagement in the cultural heritage protection. These technologies have the potential to contribute to raising consciousness of cultural heritage and to its protection at present, when many monuments are threatened by influence of time, human activities or climate change.

Chapter 1

Monuments in the landscape

Monuments in the landscape context represent significant cultural and historical points formed as a result of long continuity and development of human activity in a specific geographical space. These monuments, including architectural structures, historical artifacts and cultural symbols, are both a material manifestation of cultural heritage and a testimony of past historical periods and events. Their presence in the country is inextricably linked to the creation of local identity and aesthetic character of the region. Monuments serve not only as a visual reference to the past, but also as important places for understanding the history, traditions and values of the local community. These sites play a key role in preserving and transmitting cultural and historical knowledge to future generations.

Protection and preservation of monuments in the country involves a comprehensive approach that, along with their physical conservation and maintenance, also includes management of their use and interpretation of their significance in the present-day context. This process requires close cooperation of specialists in the field of monument care, urban planning, architecture and other relevant disciplines in order to ensure the sustainability and value continuity of these unique historical sites.

The landscape monuments are key components of cultural heritage, often representing the consequences of long-term interaction of people and their surroundings. These monuments can play an important role in preservation of historical and cultural traditions, both in their material and non-physical manifestations. Their presence in the landscape is a relevant factor of formation of the given territory identity and character.

Protection and management of these monuments requires a multidisciplinary approach that includes historical, cultural, environmental, social and economic considerations. These aspects are related to the identification, assessment, documentation, conservation, management and maintenance of monuments and their environment.

The role of the public and local communities in the landscape monuments protection is also relevant. Involving local residents and supporting their participation in the decision-making process can lead to the creation of sustainable conservation and heritage management strategies that are sensitive to local needs and interests. The monument has primarily an irreplaceable ontological value based on the relationship between man and the world, which needs to be respected and further developed. The monuments care should mean maintaining them in good condition in the context of constantly changing situations and circumstances. The monuments are cultural assets that are created and destroyed by human activity. Their existence is an integral part of people's everyday life, and be negligent about them can lead to spiritual emptiness and even to serious disasters (Stehlík 2018). Our relationship to these monuments has to be based on respect for their values. When building historical residential complexes, it is important to be aware of this asset. Absolute values, if they are depersonalized and isolated, lose their meaning and intensity, while relational values gain importance (Gogová 2014). Their protection should be understood as an awareness of their historical essence, meaning and value, along with the importance of their historical origin.

In recent years, there is a relevant trend of intensive consideration of aspects of the natural and cultural heritage care in the international conservation debate. It is essential to be aware of the effort to preserve the environment as a whole with all its specific functional contexts. Protected areas of nature have their parallel to historical sites. Arguments supporting a comprehensive view of the environment offered by interdisciplinary research on complex adaptive systems have emerged as a response to the need to connect the landscape monuments conservation with nature protection, but this unifying potential has not yet reached its full extent. It is essential to find sustainable solutions that emphasize respect for tangible and intangible heritage.

From the beginning, the conservation efforts have inevitably encountered conflicting interests, control, and restrictions on freedom of those who disagreed. In the past, individuals or groups who exercised authority and ideally had real power were more successful. Until the 19th century, the protection was aimed at preserving exceptional artefacts and life forms, except for cases such as forests, where logging, crop harvesting and game hunting were prohibited, either by the estate owners or colonial administrators towards the natives. The ecosystem preservation was more likely a by-product, since the primary goal was to deliver economic profit. Regulations on construction in the original urban centers were focused more on increasing safety and social reconciliation among neighbors than on caring for the artistic quality of the architecture or preserving its historical values. (Horáček et al. 2019, 15, 34).

Until the 18th and 19th centuries, the relationship of society to historical buildings and works of art was mostly focused on their usefulness. If the building served its purpose, it was maintained. However, if its function changed or the builder gained financial means, it was often torn down and replaced by a modern version that better suited contemporary trend. Even when old masonry or construction was used to save costs, the conversion usually preserved not only the original function, but also the style and architectural elements. The artistic value of old buildings was mostly underestimated, and until the 19th century the approach to the art of the past was influenced by aesthetic dogmatism (Gogová/Bešina/Wiedermann 2020, 30–32; Dvořák 2004). The new artistic direction often criticized works of previous generations and tried to surpass them. This attitude gradually changed the whole society and its perception of historical artistic expressions. Humanism and the Renaissance despised "barbaric" Gothic style; Classicism hated its "pompousness" and Baroque. Functionalist modernism distanced itself from the established architectural tradition even more and strove to overcome it completely (Štulc 2010, 3).

At the beginning of the 20th century, the simple views on the cultural monuments protection began to transform to the overall protection of larger urban units and the environment in which these monuments are located. In this process, the Vienna School of Art History played an important role, theoretical principles of which were accepted and further developed on the territory of Bohemia and Slovakia (Hájek 2005, 147).

The history of monument care in Slovakia goes deep into the past and reflects the rich cultural and historical heritage of this area of Central Europe. Already in the Middle Ages, there were initiatives to protect historical monuments, as churches, castles and other historical buildings were considered

FIGURE 1

important for social and religious life. In this period, some care for cultural heritage was manifested, although compared to later periods, it was less systematic (Gogová/Bešina/Wiedermann 2020, 27–29).

The flourishing Renaissance and Baroque periods brought more care for the historical monuments that was expressed in some initiatives for their preservation and aesthetic restoration. However, the modern concepts of this care began to take shape only in the 18th and 19th centuries. In this period, first institutions aimed at the protection of historical monuments were established and efforts to document and restore them were initiated (Buday 2017, Jankovič 1973).

During the 20th century, the monument care developed into the form we know today. State bodies and organizations focused on the protection of historical monuments and their systematic documentation were created (Jankovič 1981, Kutlík 1981, Orosová 1999). After the fall of the communist regime and our integration into the European Union, the monument care in Slovakia became part of the European context (Gogová/Bešina/Wiedermann 2020, 30–32). The cooperation with foreign organizations and programs aimed at the protection of cultural heritage has intensified (Dvořáková/Husová 2002; 2003 and 2004).

Understanding of a cultural object in the sense of historical record as a monument of cultural value and significance is important for elementary understanding of its own context. In the 1960s, Jakub Pavel came up with the term monumentology. We currently avoid using terms monumentics or monumentology as they can be very universal and undefined. It can leave impression that the so-called universal monument care should be introduced (Hájek 2005, 14). The meaning spectrum of the term is defined by synonyms remembrance, reminder, mind orientation, etc. If we understand a monument as a historical document, we focus on its informative value, that is, something that speaks about facts outside. Its understanding as a cultural monument emphasizes its actual existence and irreplaceable presence.

A monument is not an isolated object in the landscape; it is part of a wider environment that includes geographical, historical, cultural and aesthetic aspects. Taking these various contexts into account is essential for correct understanding and evaluation of monuments in the country and for their effective protection and conservation for future generations (Jesenský 49 - 62). We comprehend the context as a basic connection or circumstance with significance, assuming that the context is logically determined and provides a certain perception idea of the basic issue, i.e. the monuments care. More generally, the heritage care is not just a practical intervention or knowledge-based assessment of artefacts or just a science, it is a value-based process that causes intentional action. The care of monuments is a social phenomenon, effectiveness of which emerges from the essence of the activity itself, such as free human decision-making in a specific time and environment, and not only from a professional or institutional decision, which are only a means. Caring for cultural heritage is part of the basic worldview of a person who desires to be considered cultural (Jesenský 2015). Protection and preservation of pieces of the past seems to accompany human civilization since its very beginning. We were born into a culture where this care is considered an inherent part of cultural expression, which allows us to convincingly claim that this care is an integral part of our world and hence very close to us (Hájek 2005, 12). Therefore, we should not perceive the care of monuments only in a disciplinary manner as a professional matter or a social service, and certainly not as a political tool of hegemony or etatism. Theoretical and conceptual explanation of the process of contextual monuments care, which is based on facts related to the whole and enables interdisciplinary dialogue, is derived from the philosophical attitude to the cultural heritage care.

In this work, we focus on the fundamental issues of monument care, emphasizing the overall understanding of this field. Our goal is to distinguish between the intrinsic values of a monument and the external factors that affect the area. We define these external contexts as external environment of the monument care system. Boundaries between the monument conservation system and its background can be understood as communication, specifically as an attitude to a given monument value. Considering the fact that there are various partakers involved in the conservation process, who are acting and mutually interacting subjects and objects, the range of the monument care is to some extent unclear, moving and influenced by interactions between the monument and its environment. This complex interaction is conceptually illustrated in Figure 2, which visualizes the monument's relationship with its broader contextual environ. The context as an external environment does not have clearly defined boundaries and its complex structure is manifested in attempts to use it as a tool for defining the monument care system. (Jesenský 2019, 50–51).

The main components of existence of the monument care system are the subject (a human person or organization) and the object (the monument

FIGURE 2

itself). An inextricable factor of this relationship is circumstances of its realization, i.e. the context. Based on understanding of relevant different types of values, we look for the artifact characteristics that are assigned to them according to their assessment and confrontation with our internal value system. In this way, we attribute to the artifact, i.e. the monument, a certain degree of significance or value that serves as a motivation for its preservation or function. The crucial question is, to what extent the formation of the context is influenced by the internal system of the monument care and by the external environment, or to what extent are we able to control this context. So is the question of to what extent the contexts are just circumstances and to what extent they are necessary prerequisites for the system to function.

Contexts of the monument or the monument care are inseparable and forming parts of the monument itself and the care process as well, bearing traces of their time and environment. Each monument is not only an object of interest within the framework of the monument care, but it directly becomes its result, the essence of the field and the issue. Their contexts cannot simply be confused. Vít Jesenský (2019, 54) classified several criteria for the contexts classification:

- Systemic contexts - professional
- Dimensional contexts
- Contexts of human needs
- Environmental contexts

Often, we do not consider surrounding space to be the context of the monument's environment, because we almost always consider the space itself a part of the cultural heritage. Therefore, only the time dimension remains for the external context, which we have connected with social aspects. From this we can conclude that to solve the context of a given monument's environment is a community assignment first of all. Dealing with the context is not only beneficial as a theoretical determination of the area, but can also represent a potential for direct impact on the monuments care. We can see the context as a space for direct participation of the society, discussion, process of looking for consensus, team cooperation to solve problems and find partners. Contextuality of the monuments care should bring continuity in its development instead of dogmatic closure. It should serve a tool for reflection and setting parameters and thus contribute to the system effectiveness.

Exploring the context in the monuments care should not be separated from practice, as some people may struggle with this approach, and should not be limited to basic theoretical levels. Vít Jesenský (2019, 59) suggests different types of research, such as empirical, qualitative and quantitative, which are more demanding to implement. He also pointed out two objective risks: the first is the imaginary and even formal character, which can cause a feeling of incompleteness or uselessness; the second is insufficient personal authorial expertise in solving a broad and complex issue that includes several social disciplines and fields.

The concept of the monument value as a part of the subject-object relationship is firmly anchored in the foundations of theoretical considerations on the essence of the monument care and its importance. Individual value categories, as a broadly understood conceptual construct, can serve as a relative comparison of the significance that the actors of the evaluation process assign to a certain object, a set of objects or a project based on their ideological basis.

One of the monument key values is the agreed completeness. Its importance is generally recognized in both the national and international contexts of value categories. This is also applied to one of the best-known and most widespread models – the concept of so-called extraordinary universal values – which serve as a criterion for the inclusion of cultural and natural monuments on the World Heritage List (Uhlík 2019, 119).

Speaking about a monument or a urban complex completeness, we usually automatically imagine the degree of preservation of its building substance. The issue of the material wholeness of the construction is closely related to the completeness of its form. A change or absence of the idea (function/purpose) of the monument, as well as in material completeness and completeness of form, can fundamentally affect our perception of a given building or place. Since we perceive the idea of completeness in relation to the purpose of the monument, we usually take into account its primary use, for example, a church as a liturgical place, a noble manor as a residential building, or a factory as an industrial object where certain products are mass-produced. The value of the monument completeness exists in interaction with other values of monuments, and therefore acquires additional contextual meanings (Uhlík 2019, 125–127).

An extensive and significantly diverse group of the monument values includes values with a historical character. Considering the fact that these are values with memorial or symbolic meaning, the category of completeness is always crucial in their perception. However, this does not automatically mean a direct proportion between the meaning of the object, the unit and its material, shape or ideological integrity. In some cases, the fragmentary nature of the monument or the loss of some of its attributes can be a decisive factor that supports the emergence and extent of its historical value (Uhlík 2019, 128). The degree of preservation of the building substance, completeness of its shape and conceptual integrity of the monument is also important. Identification and classification of the monument into a historical period is not sufficient to determine its historical value. It is also important to evaluate its preservation in terms of historical, artistic, architectural-historical, and scientific or folklore significance (Hájek 2005, 120). The full qualification of a monument can only be achieved if the given cultural object became part of a certain historical period and witnessed events that influenced its significance. This object subsequently becomes part of the flow of history, which carries it into a new reality, where it is accepted as a symbol of the past. The value of such a monument is assessed in the context of its contribution and significance within different historical periods (Hájek 2005, 122). Figure 3 shows the structural relationships between the monument care system and its surrounding context.

If we are focused on significant works that are closely related to their material and internal structure, e.g. when studying different types of homesteads and

FIGURE 3

farm buildings, we gain sources of knowledge of historical villages and everyday life in the given area. Ambiguous conclusions regarding the role of completeness can rather appear when examining monuments as places of collective memory with their commemorative value (Uhlík 2019, 130, Jesenský 2015).

For a long time, the issue of the monument values associated with the construction and technical nature of the buildings has not been resolved. As a result, vast number of buildings and industrial sites were gradually neglected and damaged. In this category of values, there is a clear positive correlation between the degree of the object /unit preservation and its relevance as a testimony to technical progress, working methods or development in the work of prominent engineers, architects or builders. By this way, irreplaceable losses occurred concerning the outcomes of technological and industrial development (Uhlík 2019, 131).

Another field where the monument completeness is interlinked with its economic value is its indirect utility value. It is often about the ability of the monument or urban complex to function as a catalyst for the economic prosperity of the given place or region (Uhlík 2019, 133, Jesenský 2015, 57).

In the last decade, we have been observing developments characterized by postmodern relativism and socio-political changes related to the process of globalization and increasing commercialization, which are affecting the understanding of the monuments value. This shift concerns not only the emphasis on individual value categories, but also their interpretation. One of today's significant trends is increasing importance of subjective perception of the world in its diversity and manifold meanings of sights and their architectural and urban qualities. This development is well reflected in the concept of a landscape memory, where physical objects such as buildings and other architectural units may be less significant than ideas and ideological denotations to the given environment. Ideological completeness of the place, which we can be preserved only through adequate care of the region, becomes more important than the degree of preservation of the building substance and its complete form. Any attempt to restore the original mass and form of the disappeared objects or other significant interventions in the appearance of this place would have to lead to a reevaluation of our thinking about it and, consequently, to a change in its ideological integrity (Uhlík 2019, 146–147).

Regarding the category of monument completeness, an increasing trend in monuments commercialization becomes an interesting issue. Rising number of visitors to certain objects or urban units means growing pressure on these monuments. However, it can bring some benefit in their ability to generate income for their own maintenance or profit, which can act as a catalyst for the economic prosperity of a given place or region. On the other hand, there is also a growing risk of damage to their completeness, deterioration or possible destruction of structures and components, from which the buildings and their parts can suffer. It is important to find a balance between the needs of visitors and commercial use on the one hand and the preservation of historical value of the monuments on the other.

The maintenance and reconstruction of buildings protected as monuments appears to be a problematic aspect. These monuments can involve:

- national cultural monuments
- monument in the monument conservation area
- monument in the monument protection zone
- monument in the protection zone of a monument conservation area or a national cultural monument.

Monuments with the status of a UNESCO world cultural heritage site belong to special categories. Different degrees of conservation should be reflected in the different complexity of requirements set by conservationists for the implementation of construction works related to the so-called restoration of a monument-protected building (Michalík 2013, 15). Therefore, the restoration of a national cultural monument should meet the most demanding conditions, especially if it is located in a UNESCO site, while objects in a monument conservation area or in a protection zone should comply with less strict criteria. Similar approach is also expressed in the explanatory report to the Act on the Monument Fund Protection, which is important rule of interpretation. Unfortunately, in the state administration concerning the monument conservation, we often encounter an approach that does not take the above-mentioned basic classification into account. In many cases, regardless of whether it is a significant monument or just an insignificant building in a conservation zone, conservationists try to impose stricter requirements, which can relevantly increase the costs for the owner. Unlike Western and Northern European countries, this intervention in property rights is not balanced in any way, and the owner is not entitled to financial or other compensation. If the owner is not satisfied with the conditions set by the conservationists, he has several options to protect his legitimate interests. They can appeal or use extraordinary remedy or initiate a monument care inspection (Michalík 2013, 15). Builders, investors and developers rarely receive cost compensation from the state associated with construction works on a building under the cultural heritage protection. Therefore, it is understandable that they try to minimize these costs to comply with the conditions set by the conservationists and the act on the monument fund protection (sometimes they even try to find ways to avoid these requirements). First of all, their interest is to prevent problems with the state authorities that could complicate the project implementation.

The care of monuments is gradually shifting from the conservation of individual objects to the protection of entire monument areas. This trend includes a shift from the care for designated monuments and monument preserves to the protection of wider areas, including parts of the cultural landscape.

Chapter 2

Landscape as a monument

The landscape as a complex system that is formed by interaction of natural and human factors is relevant field of scientific research. It is continually developed under dynamic changes, so it is often difficult to define its current state faithfully (Chrastina 2023, 9; Otaheľ 1996, 242). The term "landscape" appeared in Renaissance Europe as soon as in the early 16th century to denote, for example, paintings of natural scenes where perspective was relevant. Since the Middle Ages, this abstract concept has developed through different stages in European countries. However, it acquired its scientific significance in the 19th century, when, for example, the works of German researchers (Alexander von Humboldt or Carl Ritter) used the term to denote both the form and physical description of the landscape. A more significant qualitative shift in landscape research occurred in the 20th century, when the term "landscape" was no longer considered an abstract phenomenon, but an empirical entity that can be subjected to detailed description and classification. Natural and cultural landscapes began to be distinguished in the 1920s, considering their objective aspects. Even today, there are some efforts among researchers to discuss the essence of the term "landscape", not only in art or architecture, but even in scientific research (Chrastina 2023, 10; Barčáková 2001, 343). This term is based on elemental experience of movement in space, moving of various objects and the flow of matter and energy in the landscape. In addition to space, the landscape is also determined by time, because spatial changes in the territory are also determined by their historical development. Thus, the landscape is a space-time with its own history, where the time aspect is simultaneously manifested as preserved traces of human activity and dynamics or time flow. One of the features determining the landscape is the spatial development of its use. Here, the assessment is often based on research into the landscape structure as a concrete manifestation of transforming nature and landscape into a cultural landscape (Chrastina 2023, 11).

Natural landscape is usually referred to as a space that was not cultivated by man before the Neolithic revolution. This landscape was formed only by natural processes. The paradox is that, for example, it does not actually exist in the territory of Slovakia because of anthropogenic pressure during the last millennium that caused its transformation into a cultural landscape. Reconstruction of such countryside at the local or regional level is less reliable (Lipský 2000, 8-10). The structure of the natural landscape potentially consists of complex geosystems that have remained untouched by man and can be analyzed according to its current biotic component and potential natural vegetation (Chrastina 2023). In present, it is not possible to talk about a natural landscape or to reconstruct it exactly, because both biotic and abiotic elements have been significantly altered by human activity (Gregorová – Hronček 2022, 7–8).

Cultural landscape is a space perceived as influenced by humans, or as a natural landscape affected by human activity. During the long-term activities of individuals or social systems, the original landscape has been transformed and enriched with new material and spiritual values. The cultural landscape can be understood as a material reality as well as a social construction that reflects activities in a particular local community. An important role in the cultural landscape research is played by its various dimensions: temporal, spatial, socio-cultural, economic, technical and ecological-environmental (Lipský 1999, 47, Žigrai 1999). The temporal and spatial dimensions are universal characteristics and can affect different areas equally. Other dimensions can be highly variable and depending on quality, intensity and clustering in different regions and over a certain time horizon. Specific character of the cultural landscape is often the result of combination of these different dimensions in particular time and space, but it can also be repeated when the cultural landscape formation reflects the respective structure of different dimensions in a given time and space (Žigrai, 2000, 230; Chrastina 2023, 12–13).

A basic characteristic of the cultural landscape is the interaction of nature and people. In this context, the landscape represents a spatial and historical phenomenon and a monument with preserved traces of human activity at the same time. Similarly, the development of human society takes place in a certain time and space, with people and society being the main actors of changes in the countryside. It is they who form the cultural landscape, which becomes the historical cultural landscape in the course of time.

The landscape is therefore a multifunctional system and space that arises as a result of natural conditions and human activity. The cultural landscape is a consequence of this interaction. It reflects the socio-economic development of a given society in a given territory with a certain time delay. The cultural landscape, as a consequence of formation of the natural environment over generations and related to the development of settlement and organization, is perceived as part of the cultural heritage (Šantrůčková 2019, 99; Kučová/Dostálek/Ehrlich/Kuča/Pacáková 2014, 13). It has been shaped over the centuries under the influence of various external and internal social, economic and cultural forces and constantly changing natural conditions. This concept comprises multifarious relationship manifestations between people and their natural environment. Cultural landscapes often reflect specific techniques of sustainable land use, which are influenced by the character and limitations of natural environment, as well as by spiritual relationship to nature (Kučová/Dostálek/Ehrlich/Kuča/Pacáková 2014, 33). Not all cultural landscapes can be described as historical. The historical cultural landscape bears significant traces of historical development and associated value aspects. It can be considered a monument that documents spiritual and material development of a given society (Šantrůčková 2019, 99).

The landscape as a monument represents a complex concept that emphasizes importance of cultural and historical heritage associated with a given territory. This approach puts accent on individual monuments as well as on their wider environmental context and relationships between them. It can bear traces of past events, cultural practices and social changes, which are reflected in its miscellaneous shapes and qualities. Protecting and conserving the landscape as a monument includes both physical interventions in the terrain and understanding and preserving its historical, cultural and environmental values for present and future generations. It is a living entity that can reflect a long-term relationship between people and their environment, way of life, traditions and values. Every aspect of the landscape – from its geographical features to architectural elements, agricultural arrangements, cultural symbols or sacred meaning - can be seen as part of a multifarious heritage that needs to be protected and preserved (Bönde Gogová 2022, 38¬39).

Identifying the values that need to be protected is the first and key step in effective protection of various qualities of a historical cultural landscape (Lowenthal 2005, 85–92). In many cases, components of historical cultural landscapes are preserved in their original form, but they are often found defunct. It is important to document disturbing elements that weaken historical value of the landscape and disrupt its composition (Šantrůčková 2019, 104–105).

Another important step in preserving the historic cultural landscape values, both in protected landscape areas and in open landscape, is the landscape monitoring. The term refers to the process of regular examining the landscape qualities in the course of time. The monitoring have to be carried out at regular basis in order to keep an eye on the landscape condition, identify components that are subject to protection, describe selected phenomena affecting the landscape quality and evaluate impacts of management decisions on the landscape under watch. The system of cultural landscapes monitoring have to be open and dynamic so it can be constantly supplemented and updated. Its structure, frequency of monitoring and types of indicators need to be flexible to respond to changing natural and socio-economic conditions

FIGURE 4

and knowledge (Salašová/Sedláček/Trpáková/Martinková Kuchynková 2015; Sedláček/Salašová/ Trpáková 2015; Šantrůčková 2019, 106).

The historical cultural landscape is a specific and complex component of the cultural heritage, where the value of time, completeness, authenticity and cultural aesthetic benefits play a key role, likewise in other monuments. It is significant material document of past social development. At the same time, however, it represents intangible evidence of social development as well. In today's urbanized society, most people do not have direct experience of living in the countryside and working in it. Consequently, the values of the cultural landscape can be appreciated and efforts to protect it can emerge, but knowledge about the landscape, especially among the lay public, is often lacking or superficial. This can lead either to indifference or to efforts to conserve the static state of the landscape. In order to preserve values of the historic cultural landscape, the monument care is confronted with numerous practical challenges regarding its management. An overall approach and awareness of the limitations in access to these landscapes is important. In their study, Karoline Daugstad and Eir Grytli (1999, 85–92) propose strategies for managing historic cultural landscapes:

•Emphasis on conservation: This approach is characterized by an effort to stop development and preserve the desired state. Here, the paradox is that the landscape is a dynamic organism subjected to permanent changes, so the absence of interventions can usually lead to natural degradation of the object of protection.

•Emphasis on education: This approach emphasizes presentation of the historical cultural landscape as a document of social development. A relevant role is played by illuminating the landscape values to the public, their educating and subsequent involving them in its protection.

•Emphasis on gentle use: This approach seeks to ensure that the historic cultural landscape remains alive. Sustainable use of the landscape is considered to be a guarantee of its preservation and development of its values.

None of the above-mentioned strategies usually exists in their pure form, but they are intermingled in practice. Even though the historical cultural landscape protection can use different approaches, it is important to define basic strategies to achieve the goals set for the field of landscape protection and to manage the measures aimed to support them (Šantrůčková 2019, 107 and 108).

The care and protection of the historical cultural landscape is a matter of public interest and its role is significant. In order to achieve this goal, it is necessary to create and develop systematic measures, to achieve coordination between different departments and to raise awareness of the discussed landscape type value (Kučová/Dostálek/ Ehrlich/Kuča/Pacáková 2014, 5).

In the international context, the cultural landscape is considered an important category of the cultural heritage. In recent decades, experts in many countries have been focusing on the protection, characteristics and diversity of cultural landscape types with increasing intensity. At the national level, it is crucial to focus on harmonizing methodologies and approaches in order to learn about the cultural landscape and to describe it. It is also necessary to unify the methods of the cultural landscape care and preservation of its value, which become the focus of inter-disciplinary dialogue.

In 1930, a law was passed in France that provided an effective framework to protect the appearance of the city and the landscape. In order to eliminate conflicts between modern construction and preservation of historical monuments, the French government enforced a strict territorial law, known as the Loi Malraux. This law represented the culmination in the process of monument care territorialization in the 20th century and opened the way for the entire cultural landscape to be perceived as a monument. The European Landscape Convention also considers the cultural landscape to be a fundamental part of European cultural heritage and a source of European identity (Hájek 2005, 148). The care of monuments is becoming more and more compact taking entire territories into account, gradually including the whole complexes of the cultural landscape, what, however, is not surprising for experts in the field of monument care. Max Dvořák (1991, 27) expressed his opinion that it is important to preserve not only the layers of the past, but also the local and historical characteristics of monuments, which cannot be modified according to certain rules, because such modifications often destroy their unique value (Hájek 2005, 148).

The European Landscape Convention (Council of Europe, European Treaty Series - No. 176) clearly stipulates that the landscape protection should be focused on the whole state territory, including diverse landscape types. However, it does not specify recommendations for areas that are home to natural or cultural value in detail. Regarding the current practice, the most

FIGURE 5

relevant documents are those created in the European context, which are part of the legal framework or methodological guidelines of professional institutions. Among these documents, the Convention on the Architectural Heritage Protection was adopted by the Council of Europe at the conference in Granada on October 3, 1985 (2001). In addition to architectural monuments and groups of objects, this convention also defines the concept of site and clearly speaks about the combination of works of man and nature.

The definition of sites, especially for the landscape historic areas, was refined by experts at meeting in Luxembourg in 1991, with the aim to define the European landscape protection more precisely as part of European architectural heritage (Heritage Landscape Sites; World Heritage List 1993). This detailed definition establishes the landscape monument areas as sites defined by their topography, with the exceptional combination of the work of man and nature, individual parts of which are interconnected by physical and spiritual ties, perceived as a whole. These sites are of particular importance for the environment and are treasured for their historical, archaeological, artistic, cultural, scientific, social or technical value. They are also exceptional due to the existence of documented relationships between different areas of interest or for their remarkable diversity within a single area of interest.

At the global level, a more precise specification of the cultural landscape values is important, which is based on cooperation of the states signing the Agreement on the Protection of World Cultural and Natural Heritage (pamiatky .sk). Although this agreement was adopted in 1972 already, its application is gradually being expanded and improved. Professional teams consisting of experts and advisory bodies develop optimized procedures for compiling the World Heritage lists and preparing documents that can serve as inspiration for other values that do not aspire to the World Heritage status. Numerous professional outputs were prepared on the basis of many international expert discussions.

Regarding the cultural landscapes, the most important conclusions were adopted at the 16th meeting of the World Heritage Committee in Santa Fé (USA) in 1992 (Operation Guidelines for the Implementation of the World Heritage Conversion, 2013), where three basic categories of cultural landscapes were defined, which were applied in practice, for example, in the Czech Republic:

1. Composite cultural landscape designed and created intentionally by man
2. Organically developed landscape, which is further examined in two subcategories:
   a) Relict or fossilized landscape
   b) Continuously developing landscape
3. Associative cultural landscape.

In order to be in compliance with internationally accepted categories, it is necessary to apply the basic categorization of the cultural landscapes of past times and the planned use of typological specifications within these categories to practice. At the international level, the European Landscape Convention is the crucial document, which is relevant for international communication and cooperation with the closest partners and which is aimed at ensuring the protection of various types of European landscape. The Convention was signed in Florence on October 20, 2000, and entered into force on October 1, 2004, after its ratification process had been completed (The Council of Europe, European Treaty Series - No. 176). Its basic definitions of the key implementation recommended are well connectible with the World Heritage Convention, because the European Landscape Convention applies to the whole territory of the signatory states, unlike the World Heritage Convention, which focuses on selected parts with specific qualities (Kučová/Dostálek/ Ehrlich/Kuča/Pacáková 2014, 14–15).

Interest in the cultural landscape is gradually becoming more focused on its inner meaning, which, however, may not be obvious at first glance. The study of associative values and relations to intangible heritage is systematically reviewed. In this way, the cultural landscape is shifting from the field of natural science research closer to the social sciences. At the ICOMOS conference held in China in 2005, special emphasis was placed on the importance of the cultural heritage environment in practical care concerning its quality, particularly within changing urban and non-built up areas. The importance of the cultural heritage environment is emphasized by the following interpretation: *"The environment of historical structures can be defined in relation to individual sites and larger units as a wider environment, in which cultural heritage of various kinds is captured, including separate buildings and arranged spaces, as well as entire historic cities and their surroundings, landscapes, cultural corridors and archaeological sites . Its importance emerges as a consequence of social, spiritual, historical, artistic, aesthetic, natural, scientific and other values of the cultural heritage. This value also impacts on the quality of the relationship associated with physical, visual, spiritual and other cultural contexts of the environment."* (Kučová/Dostálek/ Ehrlich/Kuča/Pacáková 2014, 32).

When dealing with monuments in the landscape, it is also important to consider present and future threats and challenges, such as climate change, environmental degradation, urbanization and increasing pressure on the tourist attractiveness of cities. Integrating sustainable approaches and innovative technologies into the heritage conservation and management strategies can help ensure their long-term sustainability and preservation for future generations. If the monument care is to play a role in the development of the spiritual identity of contemporary people, then it is appropriate to express concern about the possible loss of cultural values in the monument legacy. With no active public awareness and involvement into the process of preserving these values, the efforts to improve the situation in the monument care may not achieve their goal (Stehlík 2018, 25–25).

The future of the heritage protection in the landscape depends on many factors, including political interest, finance, legislation and public awareness of the cultural and historical heritage value. With growing understanding how important the conservation of cultural and historical sites is, we can expect increased efforts and investments in the protection and restoration of monuments. Considering this, increasing tourism is another relevant issue. Exploring the monument sites can contribute to the development of cultural tourism, which attracts visitors and supports economic growth in the region. This type of tourism can also help local communities improve their economic conditions and infrastructure.

Exploring of the monuments is also supported by scientific research and innovations in the field of history, archaeology, architecture and other disciplines. The new information gained by way of these researches can enrich the understanding and deepen awareness of the society of its cultural wealth.

Another important issue is the protection and conservation of the monuments. Careful examination of the monuments allows identifying endangered or damaged objects and developing plans for their protection and conservation. This contributes to sustainable development and long-term protection of the cultural heritage.

Finally, exploring the landscape monuments can also promote international cooperation and exchange of information among various countries and cultures. In this way, it contributes to building bridges among cultures and to strengthening the international relations.

Overall, exploring the landscape monuments is a significant process with many benefits that can help preserve the cultural heritage and cultural diversity in the world.

New technologies, such as digital documentation and virtual reconstructions, can contribute to better preservation and accessibility of the monuments. Cooperation of the government with local communities, non-profit organizations and international institutions is also substantial for the successful monuments protection.

It is important to realize that the protection of monuments is not a static process, but a dynamic and adaptive one. It requires continual review and updating of conservation policies and methods to ensure that the cultural and historical heritage will be preserved for future generations.

Chapter 3

Basic principles of digital methods recording anthropogenic objects and relief in the cultural landscape

From the very beginning of human settlement in the countryside, different approaches to using its individual aspects can be observed. In some regions, man persisted with minimal or no anthropogenic influence on the air and deeper layers of the soil, while gathering of food and production of primitive tools and shelters were associated with more pronounced influence on the landscape biota and relief. Human activity gradually shaped the new environment, especially in mountainous areas and on upper parts of the soil. Anthropogenic impacts on the landscape and its relief in particular, can be analyzed by means of archaeological research. Transformations or adaptations in the original natural relief, which are results of purposeful or unintentional interventions caused by human needs, are referred to as anthropogenic relief transformation. Visible anthropogenic relief forms are consequences of human activity in the landscape (Čech/Krokusová 2013, 24). The relief nature varies depending on particular sites. The human impact can be direct or indirect and is influenced by geomorphological processes. The indirect influence is manifested via the so-called anthropogenic processes, which are conditioned by local geomorphological phenomena. The direct impact is manifested through anthropogenic processes that are resulting from purposeful transformation of the earth's surface by man, using technology and effort (Falťan/Oťaľe/Gábor/Ružek 2018, 7–13). The goal of studying these processes and the relief is to understand the impact of human interaction with the landscape and to identify different forms and patterns that have arisen as a result of anthropogenic activity. In addition to this, the cultural landscape research is focused on understanding the history, values and significance of the landscape for people and supporting sustainable development, protection and conservation of cultural and natural values in the landscape area.

The cultural landscape research is an interdisciplinary process that is focused on the study of relationships of people to their environment in terms of cultural, historical, ecological and social factors. This research includes identification, documentation, interpretation and protection of diverse cultural elements that are part of the landscape.

Conventional methods of the landscape research are the important basis for understanding its characteristics and history. These traditional approaches are often confirmed by their long-term use and established practices. Field surveys, archival research, archaeological research, statistical methods, geographical mapping and ecological studies are just a few of the applied methods from various scientific fields. Cultural landscape research methods may include:

1. Archaeological research: Finding and studying archaeological artifacts and remains that reveal past societies and their relationships with the landscape (Gogová 2011, 23–24).
2. Historical research: Studying historical documents, maps, photographs and other records that shed light on past events and the landscape development.
3. Ecological research: Analyzing the relationships of humans with their environment, including the impact of human activities on ecosystems and biodiversity.
4. Social research: Investigating cultural practices, traditions, social structures and customs in relation to the landscape and its development.
5. Geographical research: Mapping and analysis of geographic data, topographic and geomorphological characteristics of the landscape.

FIGURE 6

In particular, the methods that are most relevant to our intentions and focus come to the fore:

- Historical-geographical methods
- Methods of archaeological research of the landscape (destructive and non-destructive)
- Methods of digital landscape recording

Excerption of these methods plays a key role in understanding the landscape and its use in fields such as regional planning, monument care and nature protection. Synergy of these combined methods is a basis for comprehensive understanding and evaluation of the landscape areas.

Historical geography is focused on identification and analysis of the landscape features such as villages, towns, farmland, roads and waterways, and examines their evolution and mutual relations in the course of time. It also studies the impact of historical events and social changes on the countryside and, vice versa, how the countryside can affect social and economic processes. The landscape development and changes in its structure, as well as its use over time, are examined from a historical-geographic point of view. In this discipline, methods of geography, history and archeology are combined to understand the relationship between people and their environment in the past. A number of researchers from many scientific disciplines have been and still are working on this issue (e.g. Havrlant/Buzek 1985; Miklós/Izakovičová 1997; Chrastina 2001; Chrastina 2010; Gojda 2000; Kuna 2004; Boltižiar/Olah 2009; Kolejka 2013; Hollý/Hronček 2017; Falťan/Otaheľ/ Gábor/Ružek 2018; Kučera/ Semotanová 2018; Hronček/Chrastina 2018 and many others). It is an independent interdisciplinary field of science that is connecting the space and time from the perspective of social and natural sciences (Semotanová 1998, 9–11; Rábik/Labanc/Tibenský 2013; Chrastina 2005; Tomeček 2007).

Research methods of historical geography include the analysis of historical maps, documents and archival records, iconographic sources, field surveys of human-geographical relics, archaeological excavations and studies in local traditions and cultural customs (Semotanová 1998, 25–60, Rábik/Labanc/ Tibenský 2013, 14; Kuna 2004, 237–240). Out of the great number of scientific methods and approaches, historical geography uses particularly historical methods, geographical methods, cartographic methods and methods of related sciences, e.g. botany, climatology, pedology, geology and archeology (Semotanová 1998, 61–63; Falťan/Oťaheľ/ Gábor/Ružek 2018, 24–40). Using these methods, historians and geographers can reconstruct the past landscape environment, understand its significance to past communities, and stipulate its impact on the present-day landscape condition.

Archaeological methods of the landscape research are a relevant component of the investigation of past cultures and societies by means of analysis of their physical, i.e. landscape, environment. These methods include:

1. **Cabinet advance surveys:** Archaeologists often run preliminary landscape reconnaissance using geographic information systems (GIS) to identify potential sites of prehistoric settlements, graves, artifacts, or other even more recent historical features. In addition to this, information from older excavations and published sources is synthesized and analyzed.
2. **Geophysical methods:** These methods involve the use of geophysical equipment, e.g. magnetometers, georadar or electromagnetic induction probes, to detect underground structures and artifacts without the need for destructive investigations.

3. **Non-destructive methods of the surface research:** By these methods, the surface features and artifacts that are visible on the land surface and that may include old structures, stone tombstones, ceramic fragments or other materials are examined.
4. **Destructive research (the so-called excavations):** These excavations are carried out systematically in selected sites, where archaeological artifacts are believed to be found. Excavations allow archaeologists to obtain detailed information about the structure and cultures of past inhabitants.
5. **Dating, analysis and statistical evaluation:** Once archaeological artifacts have been found, it is important to date and analyze them in order to set their age and significance for the country and society's historical past. Within these methods, there are some other ones, especially those of natural sciences, which are of fundamental importance for the landscape research.

Using archaeological methods in the landscape research, the identification, documentation and interpretation of archaeological traces and artifacts in the landscape context is realized (Renfrew/Bahn 2000, 232; Neustupný 2007, 32–42). These traces may include settlement sites, burial grounds, old roads, agricultural terraces, partially defunct structures, and other signs of human presence and activity (Gojda 2000, 150–174; Kuna 2004, 237–294). Exploring a historical landscape by means of archaeological methods is a systematic process that begins with careful planning and terrain exploration. It usually starts by identifying potential sites based on source documentation (especially maps and written sources) and field reconnaissance, while also using modern technologies, e.g. satellite imagery and geophysical scanners. Subsequently, detailed terrain mapping follows, which includes both the non-destructive recording of archaeological movable artifacts and analysis of terrain features and natural or anthropogenic elements (Kuna 1994, 19–30; Vencl 1995; Gogová/Tirpák 2014). Modern methods of geophysical measurement (Kuna 2004, 117–129), which are used to identify underground structures and artifacts, are often combined with classical methods of field and aerial observation and exploration (Gojda 2000). After thorough preparation, destructive excavation can be carried out by soundings, which are carefully planned and managed in order to preserve and document all finds (Barker 1982; Barker 1986). During these excavations, samples are collected, which are subsequently processed in laboratories using various analytical methods, such as C14 carbon dating or microscopic analysis of materials, etc. (Svobodová 1986; Renfrew/Bahn 2000, 117–150; Kuna 2004; Pollard/Batt /Stern/Young 2006; Hložek 2008; M. Jones 2011).

These methods can lead archaeologists to better understanding of past cultures and societies by studying their material culture and reconstructing their living conditions and their relations to the surrounding landscape.

At present, the cultural landscape research is dominated by modern interdisciplinary methods representing the integration of advanced technologies and analytical approaches to better understanding and documenting its geographical, ecological and cultural aspects. These methods are using new technological tools and procedures to collect, process and interpret data, thus enabling a deeper insight into the dynamics and complexity of the landscape. This complex landscape modeling combines various technological tools and analytical procedures to create detailed models comprising geological, hydrological, ecological and other factors affecting the landscape forms and functions.

The aim of these modern methods of the landscape research is to provide a detailed and comprehensive understanding of its structure, function and dynamics. They are a priceless tool for geographical research, environmental management, territorial policy planning and protection of nature and cultural heritage.

Using of modern technologies such as drones (UAV), satellite scanning, LiDAR (Laser Imaging Detection and Ranging), GPS (Global Positioning System) and geographic information systems (GIS) helps create accurate digital landscape models. They provide valuable terrain information that is used in planning and management of various projects and activities in the countryside. Considering the issue of this publication, the work will focus on the methods of digital recording and documentation of monuments and landscapes in more details.

3.1 Remote research of the Earth's surface (*Remote Sensing*)

Remote research of the Earth's surface is based on the principle of collecting and analyzing electromagnetic signals that are reflected from the Earth's surface and captured by sensors on board airplanes, drones or satellites. These signals can include visible light, infrared radiation, microwaves and radio waves, and other frequencies. These signals are recorded by sensors that generate images and data, which are then processed using various algorithms and techniques of image analysis and geographic information systems (GIS). Compared to traditional field methods, the remote sensing offers a number of advantages, among which its speed and efficiency of obtaining information is one of the main. There is no need to travel to a particular location to obtain important data about any extensive area on Earth. This method of research can collect data in different frequency spectra, including visible light, infrared radiation and radio waves (Mårtensson 2011; Lillesand/Kiefer/Chipman 2014, 241–265; Hadjimitsis/ Themistocleous/ Cuca et al. 2020; Singh 2021). In addition to this, the remote sensing provides access to remote areas that are difficult or impossible to reach physically, such as mountain ranges, jungles or polar regions. These techniques allow also vegetation monitoring and cartographic mapping. They can be applied in areas such as environmental research, land management, geology, forestry, agriculture and urban planning. These benefits make the remote sensing of the Earth's surface an important tool for better understanding and exploiting the planet resources as the surface images obtained by satellites in Earth orbit are generally available today. Satellite imagery is the process of obtaining images of the globe from artificial satellites located in space. These satellites scan the Earth's surface from different heights and in multiple spectral bands. The images are subsequently used for many purposes, such as environmental monitoring, urbanization, agriculture, geological mapping, and more. The satellite imagery process itself begins with sending a signal from the ground to the satellite, which responds by sending images back to the ground. These images are obtained from high-resolution satellites, which allow a detailed representation of the Earth's surface (Lillesand/Kiefer/Chipman 2014, 485–500). To process and analyze them, the experts use a variety of techniques and tools., including geographic information systems (GIS), image processing software tools, along with specialized algorithms for change detection, land classification, object identification, etc. (Weng 2009).

Aerial orthographic surface imaging is a technique of obtaining photographs of the Earth's surface from an aircraft in high resolution and with precise geometric correction. These images are taken using special cameras installed on airplanes that capture images from different angles and in different spectral bands.

The captured images are processed using sophisticated software tools to create an orthophoto map that is georeferenced and correct to its geometry. This means that perspective distortions and contortions are removed from photographs to provide an accurate representation of the globe´s actual surface. Aerial orthophoto images are widely used in various industries. They provide detailed and accurate information about the landscape, enabling the identification of terrain features, monitoring changes in landscape design and tracking other events and trends (Botižiar/Olah 2009, 57–65; Giri 2012, 25–37). In addition to aircraft, orthographic images of the Earth's surface are increasingly being acquired through unmanned devices such as drones.

3.2 Unmanned aerial vehicles - drones

These small, light devices are equipped with cameras and sensors that allow capturing photos and data about the Earth's surface from different angles and in different spectral bands. They are most often referred to by the abbreviation UAV (Unmanned Aerial Vehicle). It is an aircraft that is capable of flying without a human crew on board and is controlled remotely using various control and navigation systems (Themistocleous 2019, 242–243). In professional literature and legislation within the European Union, the following alternative terms are also used for these small aircrafts capable of unmanned flight:

- UAS - Unmanned aerial systems
- RPAS - Remotely piloted aircraft systems
- Drones

There are several benefits to using them in landscape mapping. First, they are able to cover even difficult-to-access and extensive areas, which is an advantage when mapping remote or inaccessible regions. In addition, they are capable of capturing detailed and high-resolution images, enabling accurate mapping and landscape analysis. Drones also allow mapping projects to run flexibly and quickly. They can be deployed as needed and promptly adapted to changes in weather or terrain. Their low operating weight and cost compared to traditional aircraft make drones an effective choice for many mapping applications. In practice, drones are used to create digital relief models, orthophoto maps, 3D mapping, landscape change monitoring, vegetation data collecting, and to perform many other tasks. Their use is increasingly widespread in geographical and environmental research, territorial planning and last but not least, in the field of cultural heritage research and documentation (Barnhart/Hottman/Marshall/Shappee 2012, 79– 83).

The importance of UAVs for research and documentation of cultural heritage can be summarized in the following aspects:

- Current affordability and quality of processed measurements
- Possibility of large-scale photography of extensive landscapes from low altitudes
- More detailed survey of monuments and human-geographic elements in the country
- More accurate documentation and recording
- Access to hard-to-reach locations
- Digitization of large architectural structures, urban units, housing estates, etc.
- Better conditions for monument reconstructions in a virtual environment.

These drones can have various types of sensors at their disposal, allowing them to collect data from the surrounding environment. There are many different types of sensors that are used for different purposes such as:

- **Optical sensors:** These sensors use light to acquire image data. They can be equipped with different types of camera sensors, such as RGB cameras (black and white, color) or multispectral cameras that can capture images in several wide spectral bands
- **Thermal camera:** They can capture infrared radiation, which makes it possible to display temperature differences in the surrounding environment. These sensors are useful for finding people or animals, detecting fires, measuring temperatures, and for other applications.

- **LiDAR:** Light Detection and Ranging is a technology that uses laser beams to create three-dimensional surface maps and obtain distance data. These sensors are highly accurate and are used to create digital terrain models, map the landscape relief, detect building heights and for other tasks.
- **Hyperspectral camera:** These cameras capture images in a large number of narrow spectral bands, providing detailed information on chemical composition and physical qualities of monuments and soils on the ground surface.

The most common are optical sensors, i.e. high-resolution cameras, allowing capturing features in the landscape or the landscape itself through two-dimensional or three-dimensional recordings. A two-dimensional landscape record usually consists of digital photographs or a map that captures the landscape elements in two dimensions - in the horizontal (x) and vertical (y) planes. These images or maps are usually created from a top-down view and provide a downward view of the landscape features. Of course, it can also be a panoramic or other lateral view of the terrain or object. A three-dimensional recording of the landscape is referred to as a 3D model or digital terrain model (DTM). These models provide the landscape view in three-dimensional space (defined by x,y,z coordinates), which allows for a better understanding of its topography and layout (Greenwood 2015). These models are created by means of various Technologies, e.g. laser scanners, stereophotogrammetry or convergent photogrammetry.

3.3 Photogrammetry

Photogrammetry is a method of obtaining accurate data about geographic or physical objects by means of analyzing photographs or images. This technique is used for mapping, terrain modeling, elevation and distance measurement, and for other applications in geographic, engineering, and environmental research (Mikhail/Bethel/ McGlone 2001). The main principle of photogrammetry is the use of geometric relations between different points on a photo or image and their actual positions in the field. These relations are derived from camera parameters, such as focal length, tilt, azimuth and pixel size, along with scanning geometry, such as camera altitude, angle of view and aircraft or drone position.

Currently, there is a wide range of photogrammetric methods that allow determining the plane or spatial coordinates of objects. These methods can be categorized according to several aspects, such as the position or number of the cameras used, their distance from the object or the method used for their processing. The output of these methods is usually 2D or 3D vector formats, spatial coordinates of points in space or orthophoto planes. In order to achieve a quality output, it is essential to have a calibrated photographic device, enabling to set the elements of the camera´s internal orientation, such as its focal length, main point coordinates, and radial and decentration distortion elements (Fig. 7.). This method works on the principle of calculating intersecting angles based on captured images and their mutual correlation (Marčiš 2019, 14–24; Brůna et al. 2015, 46–49; Brejcha et al. 2015, 39–49). The main objective is to identify location of the same points based on a specific algorithm that generates points with their coordinates. The algorithm of searching for the same points on several images from different sites uses the uniqueness of each point (pixel) and its surroundings, which makes it possible to assume that they will be found on other images as well (Marčiš 2019, 26–58).

The process of photogrammetry can be described in following basic steps:

1. **Capturing photos**: Photogrammetry begins with capturing digital images from the camera. It can be photography from the air or from the ground. In taking images, it is important to observe certain parameters, such as camera resolution, image overlay and known point references for georeferencing.
2. **Georeferencing**: The captured images need to be georeferenced, i.e. linked with exact geographical positions in field. This is achieved by

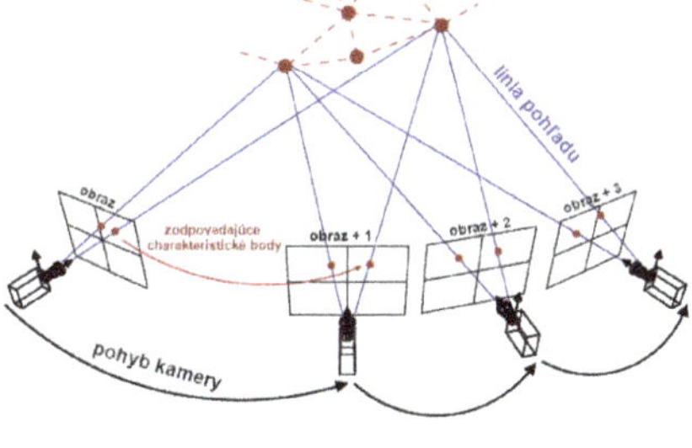

FIGURE 7

using reference points, such as GPS points or markers, which ensure that the images are in the correct positions and can be applied to create accurate maps.

3. **Orientation:** It is important to set internal and external orientation components of the camera for each image. Internal orientation involves calibrating the camera, which means obtaining its internal parameters such as lens distortion and focal length. External orientation determines the position and orientation of the camera inside the space. This is achieved by using reference points and linking them to images.
4. **Stereoscopic vision:** Photogrammetry uses the principle of stereoscopic vision, which means that it obtains 3D information about objects using paired images. Since the cameras are usually placed in various locations, each image captures the view from a different perspective. This perspective difference is used to estimate distances and create 3D models.
5. **Image processing:** After orientation and stereoscopic vision, the images are processed by applying various algorithms and techniques to extract measurable information. This includes detecting and identifying of object, measuring of distances and angles, digital terrain mapping and creating of 3D models.
6. **Final products:** Based on image processing, output products, such as digital terrain models (DTM), digital surface models (DSM), orthophotos and 3D models are created.

In photogrammetry, images or photographs are analyzed by using specialized software that evaluates the geometric relations between different points in the images and their locations on the ground. This software uses techniques, such as stereoscopic vision, image correlation and triangulation to obtain accurate data on the heights, distances and location of objects based on their images. These data are then used to create digital terrain models, 3D mapping, distance measurement and other geographic analyzes (Marčiš 2019, 78–93; Brůna et al. 2015, 47). At present, the software available for these purposes is very advanced and of high quality. This includes both commercial licensed programs and free software. In this field, the trend of constant progress leads to continuous advancement of algorithms and hence achievement of more accurate measurement results. One of the current trends is the use of so-called cloud technologies, in which a computer is used to control the program and complex calculations are handled on a remote server (Fig. 8 and 9).

To process the photogrammetric record of an effective program, sufficient RAM memory, powerful processor and graphics card in the computer are necessary. The Successful genering of the model and the process speed depend on the capacity of these components. A computer with average performance, such as any laptop, may be sufficient for creating smaller models, but the

FIGURE 8

FIGURE 9

performance requirements are much higher for capturing large-scale or complex spatial models. In creating spatial models, the final product quality depends on the number and resolution of the images used.

Methods of photogrammetric imagery differ according to the type of object to be captured. Each type of object requires a different approach when photography, whether it is an isolated solitary object, a flat linear object or an interior (Veselý 2014, 35–41). Regardless of the object type, obtaining of images with as much overlapping as possible is important to ensure better accuracy and appearance of the resulting 3D model. The minimum overlapping of images is sometimes set at 20%, but the optimum is around 60% in practice. It is critical for a credible 3D model that the viewport axes intersect in the middle of the enclosed space, as diverging axes could cause distortion or miscalculation of the 3D model geometry.

3.4 Laser scanning

In recent decades, spatial laser scanning has undergone considerable improvement. This non-contact way of obtaining spatial coordinates significantly contributes to digital modeling, visualization of building objects, structures and underground spaces. Its relevant qualities are fast data collecting, high accuracy, complexity and non-invasiveness. Numeric data are processed using specialized programs and their results are presented in the form of a point cloud. Likewise in photogrammetry, a spatial model can be subsequently generated from this data, which can be further worked with.

An important scanning device is a 3D scanner, which obtains 3D coordinates of particular area using a laser beam reflected from the scanned surface. At present, there are numerous devices working on the basis of different measurement principles (Brejcha et al. 2015, 13; Brůna et al. 2015, 6):

- Laser scanners measure the time of rapid light pulses emitted from the scanner's sensor that travel to the object and back after being reflected from its surface. They use the phase comparison method.
- Triangulation scanners can be equipped with one camera and a laser marker or two structured light cameras. These scanners scan static objects or objects rotating on a platform.

Laser scanning uses physical qualities of monochromatic light, i.e. laser, to obtain point-coordinated data in the form of a point cloud. Each point in the cloud has a precisely defined position in relation to the scanner. Number of points in the cloud is directly influenced by the image width or the capturing angle and the required resolution. When scanning larger objects, using of multiple measuring stations is essential. One location is not enough to capture an entire object in 3D. Point clouds taken from different positions are then compared and correlated to the resulting 3D model (Brejcha et al. 2015, 13–16). Before the scanning process itself, it is important to mark reference points on the object for more accurate correlation of points. The scanning process of a landscape and objects in it can be summarized in following steps:

1. **Preparation and planning:** Before starting laser scanning, it is necessary to carry out detailed preparation and planning. This includes identifying scan targets, determining optimal scan parameters, such as scan range, resolution, and scanning frequency, and evaluating ambient conditions, such as light environment, atmospheric and weather conditions. Along with this, the scanned site has to be prepared as well, including any terrain modification or removal of obstacles that could affect the quality of the scan.
2. **Scanner setup:** Subsequently, the laser scanner has to be setup correctly according to the specific project requirements and the scanned area conditions. This includes the scanner calibrating, setting the appropriate scanning height and angle, optimizing the scanning parameter (such as the beam range, velocity and intensity) and setting of sensing parameters, such as mode, accuracy and frequency.
3. **Scanning:** After successful setup, the scanning itself is carried out, which can be done from different positions and angles to cover the entire scanned area. Laser beams emitted by the scanner are reflected off the object or terrain surface. During the scanning, data about the distance, position and intensity of reflected beams are compiled to create a point cloud.
4. **Data collection:** During the scanning, the laser scanner collects thousands to millions of data points that are compiled into a point cloud, from which a three-dimensional model is created. This dataset contains

information on the distance, position and intensity of the laser beam reflected from the scanned object or terrain surface.
5. **Data processing:** After accomplished scanning, the compiled dataset is processed using specialized data points processing software. This software allows creation of a three-dimensional model of the scanned area or object, using techniques such as point data registration, noise filtering, interpolation and segmentation to achieve the resulting 3D model of high accuracy and quality.
6. **Analysis and interpretation:** The resulting three-dimensional model is then analyzed and interpreted in order to obtain relevant information about the scanned area or object. This may include identifying terrain features, measuring heights, creating digital surface models, volume models or deformation analysis. Other analyzes are also carried out, such as comparison with historical maps or models, detection of changes in the terrain or object, and evaluation of the stability and safety of structures.

Compared to photogrammetry, spatial scanning is more demanding in terms of financial and technical requirements. Data captured by expensive devices need to be compiled and processed by complex operations applying specialized computational programs, which requires a high level of skills and knowledge. As usual, cultural or scientific institutions do not have sufficient funds or personnel to cover all needs of these technologies. The limitations are mainly in staffing, time and financial possibilities (Veselý 2014, 42–47). Nevertheless, this type of scanning is used in some institutions. However, actual possibilities and benefits of using 3D digital scans in research or presentation of these institutions are the subject of debates and are not always obvious.

Scanning devices can operate on different principles with varying levels of accuracy, speed and reach. Triangulation 3D scanners measure from the end of their base. To determine position of individual points in space, they use the so-called "polar method". At each stand, horizontal and vertical angles for each measured point are recorded along with its slant distance. These distances are measured using a reflected beam of light traveling at the speed of light. Another important feature of 3D scanners is their field of view, which can vary depending on the specific device. A laser scanner includes a range finder with a built-in pulsed laser that emits flashes of infrared light, and a sensor that captures the reflected rays and measures the distance between the scanner and the object. These light pulses can reach up to 120,000 measurements per second, while some devices can capture and process over a million data per second. The latest 3D scanners are equipped with a high-resolution digital calibrated camera, which allows them to record spatial points along with real-time color information in RGB mode. In terms of their location, the scanners can be divided into aerial and ground scanners, while ground scanners are further divided into mobile and stationary ones. Both the ground and aerial scanners are similar in their structure. The measurement accuracy depends on the scanner design and on external factors, such as weather conditions. In addition, the surface of a scanned object is relevant as well, as its structure can affect the ability to reflect light back into the device, with back reflection mostly occurring in diffuse reflection. Morphology of the object has a significant impact on measurement and accurate digital geometry. As the angle of incidence increases, the intensity of reflected beam decreases towards the device. In this case, the distance measurement may be inaccurate. This inaccuracy can also be manifested when scanning the sharp edges of an object (Lenda/ Uznański/Strach/Lewińska 2016).

Last but not least, 3D scanning of monuments is a crucial tool for monitoring their condition and process of their conservation. The data obtained by scanning can be used to compare their condition in time, detect changes and deformations, and identify areas in need for repair or conservation. In this way, 3D scanning can contribute to the long-term preservation of cultural heritage in the landscape.

3.5 LiDAR

The airborne laser scanning (ALS) method uses the principles of the LiDAR (Light Detection and Ranging) method, which measures surface distance using light detection. This method works by measuring the time difference between transmitted and received reflected signal. Typically, the device is placed on an aircraft or, increasingly, on a UAV. A LiDAR device (named after the method) determines its position using a global navigation system (GNSS) with often centimeter accuracy. Based on the known position of the measuring device and the reflected and received light beam, it is possible to calculate the position of the place from where the light was reflected. Light rays are scattered into space by a rotating prism or mirror, while the number of pulses per second can reach hundreds of thousands (Carter 2012, 13–23). The result is a point cloud that is not yet classified and each point in it has its own spatial coordinates (Tab. No. 1.). By using various filtering and classification algorithms, it is possible to identify whether the light has been reflected from plants, buildings or the ground surface (Fig. 10). The point cloud that is modified and classified in this way can be further processed and visualized (Fig. 11.) as a polygonal model (Felcanová/Felcan/Lieskovský 2021, 12–14).

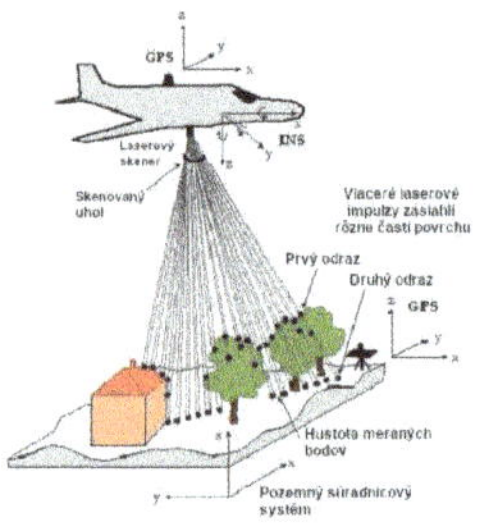

FIGURE 10

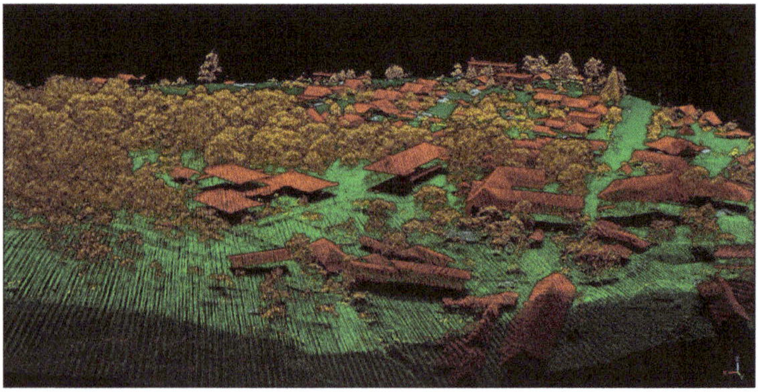

FIGURE 11

The aerial laser scanning is a key tool in the field of research into prehistoric and historic landscape structures. Characteristic features of defunct landscape structures are often not visible from the ground level and can be difficult to identify in the field due to their partial or fragmentary shape, current land use or dense vegetation. Although this method can offer significant benefits of discovering new monuments in the countryside, it is also important to verify this knowledge by confronting present-day and historical maps. When interpreting and verifying the aerial scanning results, interdisciplinary cooperation with experts in geology, pedology, hydrology and geomorphology is crucial.

In the Slovak Republic, results of aerial laser scanning (ALS) that are obtained and made available by the Office of Geodesy, Cartography and Cadastre of the Slovak Republic (ÚGKK SK), are publicly available since 2018. From 2024, the second phase of surface scanning is taking place with a density of reflected points ranging from 17 to 49 points per square meter. The main output of the ÚGKK aerial scanning is classified point clouds, which are available for downloading from the Slovak Geoportal website. In addition, it is also possible to download raster maps 5.0, which are provided

Table 1 Standard classification of points obtained by the LiDAR method according to the American Society of Photogrammetry and Remote Sensing (ASPRS)

Classification value (0:4)	Meaning (description)
0	Created, never sorted
1	Non-classified
2	Earth´s surface
3	Low vegetation
4	Medium vegetation
5	Tall vegetation
6	Buildings
7	Low point (noise)
8	Main model point (Mass point)
9	Water surface
10	Reserved for the ASPRS definition
11	Reserved for the ASPRS definition
12	Overlapping points
13-31	Reserved for the ASPRS definition

in GeoTIFF format with a resolution of 1 pixel per meter (Felcanová/Felcan/Lieskovský 2021, 13–14; Lieskovský et al. 2022). When using these materials, it is necessary to cite their source from the ÚGKK SR. Regarding the cultural heritage documentation and analysis; the use of a classified point cloud is meaningful due to its high accuracy.

Combined with geographic information systems (GIS), the LiDAR method provides a number of possibilities for the analysis and visualization of terrain and spatial data.

3.6 GIS

Geoinformation systems (GIS) can significantly influence the landscape digital outputs as they are very effective in managing, analyzing and visualizing geographic data. These systems are very helpful in acquisition, storage and integration of various types of geographic data, such as satellite imagery, aerial imagery, laser scanning, GPS data and more. With this data, complex digital models of the landscape can be developed. GIS provides tools for analyzing and processing this geographic data, making it possible to identify patterns, trends, and connections in a landscape (Ott/Swiaczny 1998). GIS enables to collect and process data on monuments, including information on their locations, architecture, condition, historical significance and conservation measures that can be further analyzed and visualized by way of digital maps and 3D models. This can help in identifying threatened areas, planning conservation work and creating strategies for sustainable monument care. It also provides tools for efficient monitoring and management of monuments by means of geographically oriented databases, which facilitate recording of changes, monitoring of condition and implementation of conservation measures.

Another significant aspect of working with GIS is creating of digital landscape models that provide detailed and accurate representations of landscape topography and can be used, for example, in infrastructure planning, flood risk analyses and terrain simulations. GIS is also useful in creation of thematic maps and visualizations displaying various geographical phenomena and landscape features. These maps can contain different layers of information for better understanding and interpretation of the landscape (Mitchell/Griffin 2021).

In general, the geoinformation system is an essential tool for creating detailed and accurate digital records of the landscape. Its ability to integrate, analyze and visualize geographic data will contribute to a better understanding of the landscape and more effective planning and management in various industries. Currently, there is a wide range of software options in the field of geographic information systems (GIS), which are constantly improving and adapting to new technological trends.

As pointed out by S. Nijhuis (2016), GIS has great potential to contribute to the development of digital culture in the field of landscape design as well. Its use offers several advantages that can have a positive impact on this area. One of the main benefits is the integration of digital tools into design processes and practice, which promotes digital literacy and awareness of the use of technology. Thanks to GIS, it is possible to create interactive maps and graphic presentations, which facilitates visual presentation and communication with clients and stakeholders through the digital environment. Another advantage is the ability to perform various analyzes and simulations. GIS allows designers to analyze various aspects of landscape design, such as terrain analysis, visibility, accessibility, and more. These analyses and simulations can help a better understanding of the environment and allow various factors to be taken into account when creating the designs. Thanks to them, it is possible to apply a scientific and analytical approach to the landscape design and create space for experimenting and innovations. These not often used aspects of GIS can subsequently be incorporated into scientifically and presentationally valuable landscape reconstructions. In addition to the above-mentioned possibilities of working in the virtual landscape space, GIS also allows simulation and visualization of virtual environments as it is possible to simulate various scenarios and parameters, such as lighting, perspective, and atmospheric conditions. In this way, realistic and interactive virtual environment can be created for exploration and survey of the reconstructed landscape.

In addition to this, GIS helps manage and monitor the landscape features and resources. In this way, it is possible to monitor changes in the landscape, collect data on its biodiversity, water resources, ecosystems and other factors. This promotes a sustainable approach to design and protection of the environment, which is essential in today's world. The role of educational and research institutions in development of digital culture in the landscape design is important as well. They provide education on GIS applications, create new methods, and study impacts of digital technologies on the landscape design.

The use of GIS brings a completely innovative approach to the application in building information modeling (BIM) and landscape information modeling (LIM) methods. Integration of GIS with BIM technologies subsequently enables creating of detailed models of buildings and structures, combining of detailed information about buildings with geographic context, and realizing reconstructions that are sensitive to historical accuracy and current needs. This approach not only supports the sustainable development and protection of cultural heritage, but also ensures that interventions in the historical environment are respecting its unique character and values.

Chapter 4

Information Modeling of Monuments and Landscapes

4.1 Building information modeling

The use of the Building Information Modeling method (BIM) and its specific extension, Historic Building Information Modeling (HBIM), represents a revolution in the field of cultural heritage conservation and management. The method is based on creation and management of digital three-dimensional models of buildings, which contain data on their geometry, structure, materials, functionality and other relevant properties. BIM is characterized by the fact that it allows the creation of integrated and interactive models that serve as a central repository for all information related to a particular monument (Diagram 1.).

BIM and HBIM enable creating of digital twins of buildings and objects, which substantially improves processes of management, documentation, restaurants and conservation of historical monuments (Liu/Willkens/Foreman 2022). This method is seen as both an opportunity and a challenge in the engineering sector, requiring considerable time, financial resources and team training, but it is an auspicious tool to improve skills and results through the application of these technologies. The main difference from traditional CAD[1] systems is that BIM is based on information-enriched parametric building elements that are combined to create entire buildings within a virtual

1 CAD is an abbreviation for "Computer-Aided Design" that are software tools used for designing, drawing and modeling various types of objects and structures in digital form. These systems enable architects, engineers, constructors, and designers to

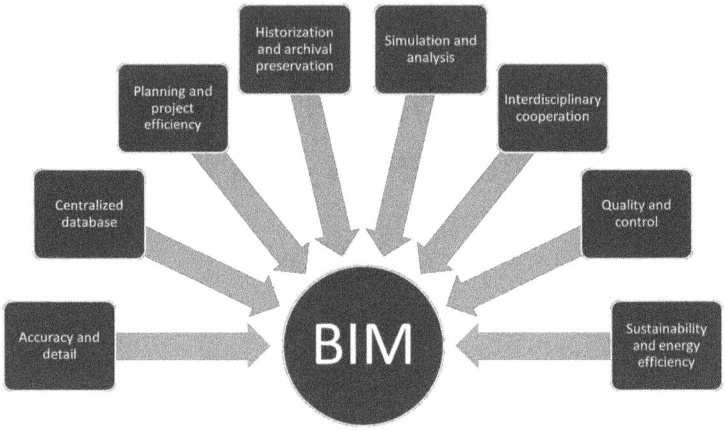

DIAGRAM 1

environment (Pitoňák 2017). It can be thought of as a hypothetical digital two-dimensional or three-dimensional model of a building that contains all relevant information and data about the building in question. This model can be visualized and edited using specialized software, and can be worked with at various stages of the building's life cycle, from design to construction to operation and maintenance. BIM gives a complete view of the construction site and ensures better cooperation and communication among all the parties involved.

HBIM refers to the comprehensive and interdisciplinary management of cultural heritage, combining aspects of geometric variables in the digital environment from 3D models to alphanumeric variables with knowledge and metadata about the object. It is a promising tool for management and documentation of historic structures, which can record relevant historical events related to the building and track the aging process of the building, but its implementation is complex and expensive (Barrile/Bernardo/Billota 2022; Laumain/Esteve Muñoz/López Sabater 2023).

---

efficiently create accurate 2D drawings or 3D models of technical drawings, products, building structures, and other design projects.

Currently, numerous heritage conservation strategies and conventions are focused on digitization of the cultural heritage objects. In some European countries, their digitization is at a high level, with the monuments being scanned and preserved in digital format. The BIM method, which creates a static model containing information about constructions, material form, etc., is a part of effective strategy intended for preservation of historical data on monuments and it has been used for a long time in some countries (Mlinkauskienė/Jankauskaitė-Jurevičienė 2019, 54). In part, BIM can also be understood as a process of cooperation aimed in production and management of structured electronic information. It is not a new technology; it has been applied in architecture, engineering and construction industry for more than twenty years and, on a global scale, is widely used mainly in the sector of new constructions (Antonopoulou/Bryan 2017, 1–2). Successful implementation of BIM, especially for large or complex projects, have to be based on a robust IT infrastructure, well-thought-out work procedures controlled by standards and protocols and (if necessary) a sustainable strategy for long-term data management (Antonopoulou/Bryan 2017, 23–27). The essential function of BIM is, of course, the creation and use of a virtual (digital) model of a building or engineering structure, which can both simulate a building construction and also improve processes of its designing, construction and use (Banfi/Barazzetti/Previtali/Roncoroni 2017, 269–273; Arayici et al. 2017; Jankauskaitė-Jurevičienė 2019, 54).

According to S. Antonopoulou and P. Bryan (2017), specific possibilities of using BIM for cultural monuments include:

1. **Historical value information management:** BIM allows integrating qualitative and quantitative information about the architectural heritage that includes visualization, development and performance of a building or site and also intangible characteristics, such as the heritage values and significance.
2. **Applications in the heritage sector:** Improved multidisciplinary cooperation and sharing of structured information can contribute to more effective management of conservation, restoration, adaptive use and repair of historic buildings and sites. BIM technology enables better spatial coordination and assessment of design options in different scenarios, which is critical for distinguished historical assets, where every

change in the construction development have to be carefully considered and justified.
3. **Use for conservation and management:** Creating a reliable knowledge base about monuments using BIM processes that can be a valuable tool for decision-making and managing processes throughout their life cycle.
4. **Archiving and information resource:** BIM can serve as an archive and information source to support future construction interventions and research and hence contributing to the cultural heritage preservation and protection for future generations.
5. **Condition management and conservation planning:** BIM permits condition monitoring, conservation planning, preventive maintenance and intervention management at both strategic and operational levels.
6. **Heritage interpretation and management:** Use for heritage interpretation, visitor management and evaluation of possibilities for intervention into monuments.

The use of BIM (HBIM) in the cultural heritage sector brings relevant benefits in terms of streamlining processes, better planning, cost reduction and improving the carbon footprint of historic buildings and sites. At the same time, it enables to coordinate and evaluate the proposals under different scenarios in better way, which is essential in the cultural heritage protection and preservation of historical values (Antonopoulou /Bryan 2017). Applying the BIM (HBIM) methods can be of substantial profit in accurate reducing of design and construction costs, eliminating errors and conflicts among various building elements, managing effective time and costs, or in high-quality and timely communication among the construction project participants (designers, manufacturers, builders, appraisers, etc.) and in many others (Mlinkauskienė/Jankauskaitė-Jurevičienė 2019, 56).

HBIM is a relatively new branch of academic research and does not yet seem to have received adequate attention from experts in the field of heritage management and conservation and, even worse, nor in academic circles. In the process of creating HBIM of a monument, digital representation of all its components is captured, which can technically be defined as object-oriented parametric modeling. HBIM parametric objects are representing various architectural elements, structural elements, systems, other components and spaces and are intelligent carrier of information. Therefore, the HBIM method has a very wide application in the field of research, management and presentation of cultural monuments, where it brings considerable benefits for the conservation, management and research of historical buildings and landscapes. This method offers a solid framework for a multidisciplinary collaborative process of information production and exchange, which results in creating of a reliable shared information resource that is used as a basis for decision-making, communication, planning and consultations. HBIM offers a framework for better information management by integrating digital technologies and collaborative work processes applied in different sectors of the construction industry using common standards (Antonopoulou/Bryan 2017, 2). In accurate digital models of monuments, heterogeneous data including historical photographs, oral traditions and even music, are integrated into a unified model (Fai et al. 2011). In this way these technologies can open up new possibilities for the research, preservation and presentation of historical objects. In modern architectural designs and infrastructure, BIM has shown potential benefits in the phases of designing and construction, with significant gains during the operational phase. These benefits can also be applied to projects in the heritage care sector, such as conservation, reconstruction, adaptive reuse, extension and repair (Antonopoulou/ Bryan 2017, 3–4).

The HBIM method has wide possibilities of use in practice, not only for preservationists, but also for monument owners or custodians. Alike in the new planned construction sector, potential historic building conservation projects, in which HBIM is implemented, can bring benefits in terms of capital and operating expenses (Antonopoulou/Bryan 2017, 2). The monument care projects typically depend on multidisciplinary cooperation, with experts and specialists from different disciplines exchanging complex information and data about a monument to interpret its value and significance. This understanding is crucial when making decisions concerning future interventions, conservation and management of the object. The research of every cultural monument and assessment of its heritage values and significance should be done to ensure that decisions leading to its change are conversant and made with deep understanding to all these aspects (Jordan-Palomar et al. 2018).

Of course, it is essential to define the requirements clearly and subsequently develop the strategy of BIM delivery. The clearly defined BIM strategy is crucial for pilot projects and case studies. The delivery method effectiveness can be evaluated by comparing actual and expected results (Hardin/McCoole 2016, 45).

By introducing collaborative BIM practices, working relations among various participants in construction projects are changing with significant legal implications. Existing contracts often do not cover the various relationships affected by BIM implementation. The clearly defined BIM strategy should not be seen as just another contractual obligation or formality. To create a plan for information delivery using BIM is good professional practice for both the construction and management sector and for the research and heritage exploration projects (Antonopoulou/Bryan 2017, 32). However, there are challenges in implementing BIM and HBIM in practice, such as high costs on data collection and modeling, lack of standardized procedures, and need for specialized software and hardware. Despite these challenges, HBIM is considered a promising tool that needs to be gradually developed and put into practice in the favour of future cultural heritage protection and management (Liu/Willkens/Foreman 2022).

The basic starting point of the HBIM process is a thorough knowledge of the existing building, including the materials used. Representation of existing geometry in a 3D BIM environment also requires detailed and accurate knowledge of the historic building physical shape and its architectural extensions. Historic BIM should be based on accurate "as-it-is" metric data sets (preferably in 3D space) that are documenting the position, size and dimensions of all visible surfaces, components and the context of the historic property, referenced to a local or global coordinate system. Additional research, quality surveys, cabinet reviews and site inspections may be required, depending on the project scope and the requirements of the property manager or owner. Non-geometric data acquisition may include condition surveys, archaeological excavations, material analyses and other surveys (such as M&E[2] systems) carried out by relevant specialists. The amount and quality of information available about the monument historic nature can vary significantly, depending on the building's type, previous use, ownership and management systems. Existing information may be present in physical form (drawings, etc.) or in electronic format on-site or

2 M&E systems, where "M&E" stands for "Mechanical and Electrical", are part of engineering systems and building infrastructures that include all mechanical and electrical components and services. These systems are essential for the safe, efficient and comfortable operation of a modern building or structure and include a wide range of technologies and equipment.

scattered across several online resources (Cloud systems) and potentially be subject to intellectual property or security restrictions. The use of original historical knowledge in the HBIM project is mainly determined by two qualities: its availability and validity, i.e. relevance (Antonopoulou/Bryan 2017, 10–12).

According to Dore and Murphy (2012), the aspects of data collection and processing for proper compliance with HBIM should follow two principles:

- **Data Collection:** Data collection for HBIM involves the use of laser scanning and photogrammetry, which make it possible to record very high and accurate field details for monuments documenting. This process can be automated and accelerated by means of precise parametric objects that can be adapted and mapped to the survey data (Dore and Murphy, 2012; Valero et al. 2017).
- **Modeling and information management:** After creating a 3D model, the model is integrated into 3D in a program that supports the GIS environment for further analysis. It uses the international 3D urban modeling framework CityGML[3], which provides an interoperable framework for modeling 3D geometries, semantics, topology, and appearance properties. This process is aimed at bridging the gap between parametric CAD modeling and 3D GIS, taking advantage of both systems to aid in the heritage site documentation and analysis (Dore/Murphy, 2012).
- The virtual collected data modeling and processing in BIM (HBIM) begins with creating of a monument digital 3D model that contains detailed geometric and material information about its each construction component. Approaches to virtual HBIM modeling, focused on the specifics of historical buildings modeling, can be summarized as follows (Antonopoulou/Bryan 2017, 14–20):

3 CityGML (City Geography Markup Language) is an international open standard for 3D modeling of urban objects that was developed and is maintained by the Open Geospatial Consortium (OGC). The standard defines a common semantic information model structure and exchange format for storing and transferring digital 3D models of cities and landscapes, including buildings, terrain, roads, waterways and other urban features.

- **Scan-to-BIM approaches:** These approaches use 3D scanning, such as laser scanning or photogrammetry, to create pinpoint digital representations of existing structures. The data scanned are the basis for creating detailed 3D models in BIM software. This method is particularly useful for capturing the complex geometries and details of historic buildings.
- **Parametric modeling and custom content:** For unique historical components that are not a standard part of BIM software libraries, it is necessary to create custom BIM objects. This approach can provide the user with more accurate representation of specific characteristics and details of historic buildings, but it can require weighty work and specialized knowledge.
- **Use of existing 2D and 3D CAD data:** In some cases, existing 2D or 3D CAD data can be used as a basis for creating a BIM model. This approach can be effective if accurate and up-to-date CAD drawings are available, but it may require additional adjustments to make the data compatible with BIM processes.
- **Hybrid modeling:** This approach combines various methods, such as scan-to-BIM, parametric modeling, and the use of existing CAD data, to create a complex BIM model. This modeling makes it possible to take advantage of the benefits of each method while minimizing their limitations, which is ideal for projects targeting complex historic buildings.
- **Level of detail (LOD[4]):** The discussion also highlights the importance of determining the appropriate level of detail for different parts of the BIM model to ensure the right balance between the model accuracy and software performance.

In creating and modeling BIM (HBIM) models, BIM software libraries play a key role. These are very important collections of digital objects and components that can be used in modeling. They contain predefined models of

4 Level of Detail (LOD) in the BIM (Building Information Modeling) context refers to the degree of accuracy and detail, with which a digital model of a building or part of it is developed. LOD is divided into different levels, with each level representing a higher degree of detail and information about the modeled object. As the LOD number increases, from LOD 100 to LOD 500, the amount of geometric detail and related information (e.g. materials, qualities, technical specifications) in the model increases.

building elements, such as windows, doors, ceilings, furniture, installation elements and many others, which are specified in detail, including their qualities, materials, dimensions and technical specifications.

The benefit of using BIM software libraries is that architects, designers and engineers can quickly add and integrate precise and standardized components into their digital models, which is simplifying the designing process and increasing its accuracy. These libraries also support more effective cooperation of the project participants by providing a consistent and uniform basis for information exchange. They can be specific for individual manufacturers and thus providing detailed information about particular products, or they can be created as generic libraries that provide standardized solutions for common construction components. These digital resources play a key role in the BIM process, because using them, some design aspects of the project can be automated and the quality and coordination of its documentation can be increased.

When working with historic buildings in a BIM environment, the users may face specific challenges that can arise from the software oriented to modern building standards and materials. As usual in these libraries, there is often a lack of specific elements or materials needed for authentic modeling of historical objects, such as traditional building techniques and materials, which may require creating your own models from scratch. This need to create models by hand is time-consuming and emphasizes the uniqueness and complexity of historic buildings, which contain architectural features and details difficult to replicate using standardized objects. In addition, in order to preserve historical and cultural values, it is essential that the models are as accurate and authentic as possible, which is not always possible with standardized models in the libraries. Integration of modern technologies and systems into existing historical structures poses an additional challenge, as it is necessary to find objects that would fit into the historical context while meeting present-day standards (Antonopoulou/Bryan 2017, 18). The solution can be found in creating of specialized BIM libraries, adapting of already existing objects or close cooperation with experts in historic buildings and conservators to ensure accuracy and authenticity of the models created.

Both BIM and HBIM go beyond the traditional concept of 3D modeling or a specific type of software; it is a process that can facilitate the information exchange and management among various disciplines involved in the project. One of the basic requirements of BIM to increase efficiency is that information is not duplicated, but is shared and reused within the project team. The interdisciplinary cooperation thus depends on interoperability and meaningful data exchange.

It is generally accepted that no single BIM tool can cover every requirement of design, construction and process management. Instead, project teams have to create more or less complex workflows that involve several interconnected software. Different specialists should also be able to choose the digital tools best suited for their task, regardless of the software choices of other team members (Antonopoulou/Bryan 2017, 18–19; Barrile/Bernardo/Billota 2022).

Implementation of HBIM requires integrating of different types of information and guaranteeing their high quality, which needs development of a web archiving system. This system enables management of structures (tangible value) supported by new technologies and sharing of knowledge (intangible value). Creating of 3D digital models in the BIM environment is a relevant step that is one of the most important parts of the process. The challenge, however, is to create parametric objects that are not included in BIM libraries, which requires considerable modeling time, or they can be found in catalogues of distribution companies. A great effort is needed to reproduce the geometric uniqueness of monumental buildings (Dore/Murphy 2012). The BIM and HBIM implementation requires specific software solutions. For example, Autodesk Revit was chosen for the development of HBIM models because of its ability to integrate geometric and material properties of historical buildings (Rodríguez-Moreno et al., 2018, 5). The challenge, however, remains as Revit and other software may not be fully compatible with historical features, such as patina, structural and aesthetic qualities linked with particular settlement or historical renovations (Liu/Willkens/Foreman 2022). The main challenges in data collecting and modeling process are large volumes of point clouds and high demands on computing power, which require development of more efficient software algorithms and automation of modeling. For example, the Rhino Grasshopper-ArchiCAD combination offers a new workflow for automatically converting point clouds from TLS or SFM into textured 3D grids and thus BIM objects that can be included in an

HBIM project (Dore/Murphy 2012, 374–376). It is also important to use software such as Graphisoft ArchiCAD, which supports Geometric Description Language (GDL) for creating parametric objects (Dore/Murphy 2012, 375; Fridrich/Kubečka 2014). Integrating of the programming environment under the Geometric Description Language (GDL) allows creating of parametric objects that are the basis for 3D geometries such as walls, slabs, floors, columns and stairs. These parametric objects allow creating of geometries that are not adaptable to the morphological irregularities that sometimes occur on the walls and columns of historic buildings. The remote sensing techniques such, as terrestrial laser scanning (TLS) and photogrammetry with the SfM algorithm are becoming essential to solve this problem (Bagnolo et al. 2019).

Databases in the context of BIM and HBIM methods serve as a central repository system for all information related to a given project. They are substantial for the efficient management, documentation and preservation of buildings, whether they are new buildings or historic ones. Converting a model to CityGML and integrating it into GIS platforms, such as ArcGIS, allows for further analysis of geometry, semantics, attributes, external links and topology. This information may contain historical data on the cultural object origin and chronology. Furthermore, historical documents can be linked to the model components for complete documentation. The model can also be used for tourism applications and easily shared over the web for its visualization (Dore and Murphy 2012). At the same time, these models can support strategic decision-making regarding the protection, conservation and presentation of cultural heritage (Rodríguez-Moreno et al. 2016; Castellano-Román/Pinto 2019).

The importance of databases in BIM and HBIM can be summarized in the following points:

1. **Centralized information management:** Databases allow all the information associated with a building to be collected, stored, and managed in one place. This includes geometric data, material specifications, performance data, and historical records, making it easy to access and update information for all project participants (Fai et al. 2011)
2. **Cooperation support:** Databases in BIM and HBIM enable effective cooperation of the project team members, including architects, engineers,

builders and conservators. The real-time information sharing helps reduce misunderstandings and conflicts, thus improving project coordination and efficiency (Quattrini et al. 2017).

3. **Conservation and documentation of historical objects:** For HBIM, databases are critically important for conservation and documentation of historical objects. They allow preserving of detailed records of architectural elements, materials, construction techniques and history of objects, which is invaluable for conservation purposes and research (Oreni et al. 2014).
4. **Integration of heterogeneous data:** HBIM databases can support integration of various types of data, including tables, graphs, images, texts and links. This capability provides a comprehensive representation of a building or historical object that reflects its physical, historical and cultural aspects (Liu/Willkens/Foreman 2022, 52–54).

The tasks arising in the process of HBIM databases creating can be diverse and involve technical, methodological and organizational challenges. One of the main challenges is integration and coordination of different types of data obtained by various methods, including laser scanning, photogrammetry and historical research. This requires a high level of expertise and multidisciplinary cooperation of architects, conservators, historians and IT experts. Moreover, processing and managing large volumes of data can be demanding in terms of computing capacity and storage (Rodríguez-Moreno et al. 2018; Fadli and AlSaeed 2019). Another relevant problem is the preservation of semantic information and its integration into the model. This includes both the geometric details and historical and archaeological information, which is essential for the monument understanding and interpretation. Ensuring that this information is accurately and reliably integrated into HBIM models requires detailed knowledge of historical and conservation principles as well as advanced modeling skills (Rodríguez-Moreno et al. 2018).

Last but not least, problems can also arise in relation to the level of detail (LOD) required for various components of HBIM models. Finding a balance between the need for high detail in certain parts of the model and the need to limit the overall complexity of the model to a sustainable level is a challenge that needs careful planning and decision-making (Fai et al. 2011; Ham et al.

2020). These findings show that the successful implementation of BIM and HBIM requires both the sophisticated software tools and also well-structured databases capable of processing and storing large amounts of data.

Resulting from the detailed analysis, implementation of BIM and HBIM in the field of cultural heritage protection is a complex process demanding interdisciplinary cooperation, specific software solutions and effective data management. Despite the existing challenges, modern technologies such as 3D scanning, parametric modeling and advanced software platforms offer significant potential for more accurate documentation, analysis and conservation of historical objects.

4.2 Landscape Information Modeling

Landscape Information Modeling (LIM) is becoming an integral part of landscape architecture and environmental planning, as BIM technologies and methods spread deeper into architecture, engineering and construction. Essentially, LIM is a digital representation of the landscape environment that embraces physical, biological, ecological and other characteristics of the landscape. These digital models can be created using multifarious software tools and technologies, such as geographic information systems (GIS), 3D modeling, and simulation. LIM can be understood as a comprehensive digital model that captures the landscape diversity, including terrain, water, vegetation, infrastructure and other components. This model can incorporate various layers of information that allow different landscape aspects to be analyzed and visualized (Diagram 2).

For example, there can be layers for topography, land use, cultural heritage, ecological characteristics, and more. The term HLIM, which stands for Heritage Landscape Information Model, is defined as an integrated information framework that is supposed to support the practices of conservation of the landscape cultural heritage. HLIM aims to facilitate the digital management of the landscape cultural heritage information through the use of modern technologies, such as virtual reality, landscape monitoring tools and special information platforms for cultural and historical information. HLIM extends the concept of BIM to the field of landscape conservation, emphasizing the need of both the physical landscape components and intangible aspects, such as cultural experiences and pieces of history associated with specific

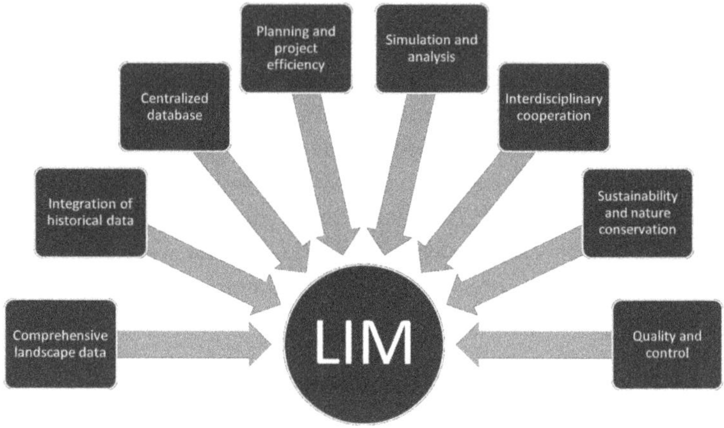

DIAGRAM 2

sites, to be incorporated into the model (Yang et al. 2019). In this context, HLIM aims to capture, analyze and visualize a wide range of landscape data for better understanding and protection of its cultural and historical values.

The LIM method itself enables landscapers, architects and other participants to better understand the landscape environment, to anticipate its development and propose effective interventions or projects with respect to various landscape aspects and its sustainability. The method also offers better cooperation of different experts and partners involved in the landscape planning and management.

The work of A. Mohammad A. A. Aliyu (2023) emphasizes the need of identifying the BIM importance in the landscape architecture in order to promote deeper adoption of this method. This presents a challenge and opportunity for the landscape architects to improve their practice through effective and efficient use of collaborative BIM technology (Ahmad/Aliyu 2023, 531–532). The advancement in the landscape architecture, urban planning and open space design can be seen in transition from a single discipline to several independent disciplines at technical, disciplinary and scientific levels. The landscape is presented as a structural component of reality, comprising a set of elements that form a complex system. This system is characterized by subjective perception and interaction with the observer. Currently, various types of the landscape - original natural, anthropogenic, mixed and also archaeological - are being intensively discussed, pointing out the complexity of the landscape project, which reflects the combination of anthropic and natural elements. The main goal of experts in this field is to analyze the current state of digital representation of open space projects and formulate a methodology for integrated work and representation of particular components. This approach emphasizes the need for interaction of experts from different professions involved in open space projects, regarding both the new transformations and maintenance and documentation of the existing situation. The research identifies LIM as an application of information systems to an open space, not an architectural product, i.e. it reflects its adaptation to new technologies and legislation focused on BIM and public procurement (Cianci/Molinari 2019; Forte 2006).

Currently, a deeper analysis of the purposes and forms of modeling in the landscape architecture and environmental planning is practicable, as there are ongoing discussions about various purposes for which landscape models are created - from simulation and explanation to experimentation and communication. It is pointed out that models can be concrete or abstract, simple or complex, and it is emphasized that current digital models are increasingly used to simulate and explain natural ecological processes. The ability of these models to capture both the static and also dynamic aspects of the landscape is significant and opening up new possibilities for the landscape architects and scientists to better understand and communicate the landscape processes (Ervin 2023, 1–14).

The historical development of LIM reveals a remarkable journey from traditional modeling techniques to present-day digital approaches. This evolution mirrors broader advances in technologies and information systems that are transforming the landscape architecture and environmental planning. In its early stages, the landscape architects and planners primarily relied on hand-drawn designs and physical models to represent the landscape designs and plans. These traditional methods, although effective to some extent, had limitations in their accuracy and ability to represent dynamic processes and interactions in the countryside. Developing digital technologies together with emerging geographic information systems (GIS), 3D modeling and simulation techniques have brought new opportunities for the landscape architects. These tools provides them with more accuracy in simulation and visualization of environmental processes, visibility analyses, hydrological models and other key aspects of the landscape planning and design (Ahmad/ Aliyu 2023, 531–532). Professionals in the landscape research and landscape planning use models in different media and at different levels of abstraction for various purposes. This approach points out to the transition from analog models, such as cardboard and polystyrene models, aerial photos and GIS data layers to digital 3D models and simulations. This shift to digital tools has enabled better integration of different types of data, leading to a multidisciplinary approach in the landscape architecture. Modern computer graphics tools have made it possible for the models of both scientists and designers to be fully spatial and dynamic, increasing their ability to be evaluated in very different ways (Ervin 2023, 1–14). This evolution has improved accuracy and efficiency in the landscape design and planning and at the same time enabled the landscape architects to better understand and communicate dynamic processes and interactions that shape our countryside. Thus, the advances in digital technologies and information systems have opened new horizons for innovation and collaboration in the landscape architecture and environmental planning.

The Landscape Information Modeling (LIM) presents a range of benefits and challenges that have evolved over its historical context, from traditional modeling techniques to present-day digital approaches. These improvements have increased efficiency and effectiveness in the landscape architecture and environmental planning, but simultaneously have brought new challenges for the professionals to be faced with.

The benefits of the landscape information modeling are strongly reflected in the possibilities that digital tools offer for the simulation, visualization and analysis of the landscape projects. Ahmad and Aliyu (2023) point out the potential of BIM to overcome some limitations in the design, construction and management of the built environment. BIM enables the landscape architects to use collaborative technologies effectively and efficiently that improves planning, organization and communication of various project participants (Ahmad and Aliyu 2023, 531–532). The possibility of sharing a landscape model among all users in the process of design, construction and operation of architectural objects or structures is a very important aspect of LBIM technology. This possibility is ensured by the use of a standardized data model IFC (Industry Foundation Classes), which is focused on describing architectural, construction and infrastructure data. IFC, as an open international standard ISO 16739-1:2018, allows independence from software providers and so far has a limited number of IFC classes in the field of landscape architecture and land management, which points to the need of expanding its capabilities to better cover the landscape architecture needs (Borkowski/Wyszomirski 2021). The tools of digital modeling allow professionals to integrate different types of data and take a multidisciplinary approach. This progress leads to a better understanding and communicating of the landscape dynamic processes and offers high-leveled and competent innovative designs (Ervin 2023, 1–14).

Promising software solutions and technologies for LIM, whose full potential for the landscape architecture, planning, management and sustainable development has not yet been fully realized, provide innovative opportunities for improving the process efficiency, data accuracy and for their integration with other software tools. The need of expanding the already existing BIM tools and developing specific LIM applications underlines the importance of further research and development in this area. The LIM integration into a wider scope of planning and conservation projects can offer a promising heading towards improvement in the landscape cultural heritage management and preservation.

The need for the LIM and BIM integration is discussed by A. S. Borkowski and M. Wyszomirský (2021) in their work "Landscape Information Modeling: An Important Aspect of BIM Modeling". The authors identify several software solutions used in BIM that could be applied or extended for LIM needs, including Vectorworks Landmark, Land F/X, LandCAD, Siteworks, Archi Terra, AutoCAD Civil 3D, Autodesk's Revit, and Graphisoft ArchiCAD. These softwares, although originally designed to be applied in architecture and engineering, can provide the basis for development of specific tools for LIM due to their ability to model both physical and intangible landscape aspects. Borkowski and Wyszomirski also emphasize that the current IFC (Industry Foundation Classes) model, which serves as a standard for the exchange of BIM data, does not allow incorporating the data about the landscape infrastructure and objects. In the work, the necessity for expanding the IFC model to include these objects is emphasized that would significantly improve the integration and exchange of data between BIM and LIM systems.

Some of these tools and their potential benefits in the work of landscape architects are listed here:

1. **Vectorworks Landmark:** Offers comprehensive design, drawing and documentation tools specifically tailored to the needs of landscape architects. It includes functions for terrain modeling, plant placement and site planning, making it easy to plan and visualize the landscape in detail.
2. **Land F/X**: Integrated with AutoCAD, Land F/X specializes in irrigation planning, planting design and site planning. Its database approach enables efficient management of plant and equipment data, simplifying the landscape design process for professionals.
3. **LandCAD:** Another CAD solution providing tools for site planning and landscape design. Its functionality includes terrain modeling and landscape visualization, which helps architects create detailed outdoor spaces.
4. **Siteworks for Revit:** Enables to plan and design the site components in the Revit environment at an early stage. It offers the possibility to model terrain, roads and sidewalks that are seamlessly integrated with building models.
5. **Archi Terra:** An extension for ArchiCAD focused on terrain modeling and landscape design. Provides landscape architects with tools to create detailed terrain models, including slopes, hills, and excavation areas.
6. **AutoCAD Civil 3D:** Civil engineering and documentation software that also serves landscape architects in the design, analysis, and documentation of civil projects. Its functions include terrain modeling, storm water management, and civil structures, which is beneficial for the landscape and site complex planning.
7. **Autodesk's Revit:** Although primary architectural design software, its BIM capabilities enable to integrate the landscape elements into construction projects. Landscape architects can use Revit for site planning,

terrain modeling, and integration of landscape features with architectural designs.

8. **Graphisoft ArchiCAD:** Offers great BIM capabilities with applications in the landscape architecture, especially in the design and documentation phases. Its modeling tools can be used to represent terrain and integrate landscape elements into architectural projects.

Each of these software solutions provides specific features that can improve the workflow for the landscape researchers. Starting with detailed terrain modeling and vegetation management up to complex planning of a site and its integration with architectural models, these tools offer a range of features to meet diverse needs of the landscape design and planning. The integration of LIM into these software tools signals a shift towards more complex and integrated approaches to design, enabling the landscape architects to use digital technologies for sustainable, efficient and visually appealing landscape solutions (Borkowski/Wyszomirski 2021, 7–22).

However, the challenges of landscape information modeling include the need for specific software for the landscape architecture and mastering new technologies. Ahmad and Aliyu (2023) identify the absence of specific software for "landscapers" as a major obstacle to BIM adoption in the landscape architecture, which can hinder cooperation with other BIM-using professionals. This situation arouses the demand for the competent to group together and put pressure on the software vendors to realize the importance and potential of producing specific LIM software (Ahmad and Aliyu 2023, 531–532). Ervin (2023) further points out to the challenge involved in the translation between the "source" world and the "target" representational medium, emphasizing the need to understand both factual and procedural knowledge in the modeling process. This multidisciplinarity requires a wide scale of knowledge, from explicit scientific to individual or practical one, which can be a challenge in the context of increasing complexity of modeling media (Ervin 2023, 1–14). It is necessary to adapt the landscape project to new technologies and overcome classic work patterns. Better understanding, analysis and prediction of the evolution of urban structures and natural phenomena by means of BIM and GIS platforms are emphasized. This can prove the integrated digital databases of architectural artefacts and natural systems to be necessary for systematic study and planning of maintenance interventions (Cianci and Molinari 2019).

Despite the challenges related to the adaptation and integration of landscape information modeling, the benefits such as increased efficiency, better decision-making and innovative designs far outweigh the obstacles. Future development in this field calls for continuous education, collaboration and innovation among the landscape architects and other AEC[5] professionals in order to maximize LIM potential for sustainable development of the landscape environment.

The use of the LIM and related HLIM concept brings new possibilities for landscape virtual visualizations and simulations. Extending the capabilities of traditional Building Information Modeling (BIM) into the field of cultural landscape conservation and presentation, HLIM enables to create detailed 3D landscape models. These models not only include physical elements, such as terrain, vegetation and water features, but also allow for the integration of cultural and historical structures.

The HLIM approach to virtual reality and monitoring tools suggests possibilities for creating interactive 3D models and virtual tours that provide immersive experience of cultural landscapes without the need for physical presence. This aspect of HLIM has a great potential for example in education, tourism and cultural heritage conservation, allowing the users to experience a cultural landscape virtually, while providing tools for their conservation and sustainable management.

Furthermore, HLIM supports the simulation of environmental processes, which opens up possibilities for understanding the landscape development and predicting its future changes. These simulations, including for example erosion, plant growth or land use changes, are invaluable for planning the landscape conservation and its sustainable development. By using HLIM as

5 The field of AEC in the LIM (Landscape Information Modeling) context refers to the interdisciplinary cooperation between architecture, engineering and construction. The term is often used in discussions of BIM (Building Information Modeling), where it emphasizes the integration and coordination among various professions and project phases, from design to construction and maintenance of buildings and other construction projects.

a tool for education and awareness, it is possible to raise awareness of the landscape cultural heritage value and support its protection. Interactive visualizations and virtual landscape models can serve as an effective means of presenting cultural landscapes to the public and hence increasing its engagement and support for their preservation.

Finally, digital terrain records created by means of HLIM can serve as a basis for their restoration or a resource for scientific research, thus ensuring the cultural heritage to be protected and accessible for future generations. Hence, this complex approach to the landscape information modeling through HLIM opens up a wide range of possibilities for visualizing, simulating and protecting the cultural landscape heritage (Yang et al. 2019).

The landscape information modeling (HLIM) potential in the cultural heritage conservation is great and can bring many advantages and benefits:

- **Digital documentation and the heritage conservation:** LIM can serve as a tool for the cultural heritage digital documentation and conservation. These models can capture details of historic sites, buildings, monuments and landscape features with high accuracy, allowing them to be better protected and preserved for future generations.
- **Conservation analysis and planning:** Using LIM, the professionals involved can perform various analyses and assessments, such as identify endangered areas, assess the monument conditions and propose measures for its protection and sustainable restoration.
- **Visualization and communication:** LIM allows creating realistic visualizations of historical sites and monuments, which helps the wider public to better understand their cultural and historical significance. These visualizations can also serve as educational tools and promote its awareness of the cultural heritage.
- **Integration with GIS and geodatabases:** LIM can be integrated with geographic information systems (GIS) and geodatabases, allowing better management and analysis of the cultural heritage in the context of its geographic location and interactions with the surrounding environment.
- **Supporting the decision-making process:** Based on the LIM data and analysis, the authorities in the monument care can make decisions regarding the cultural heritage protection and management more effectively.

This information can be used in policy making, project planning and decision-making processes.

Application of landscape information modeling (LIM) in reconstruction of historical landscape monuments in a virtual environment represents a fascinating breakthrough in the field of archaeology, historical geography and cultural heritage conservation. This approach enables detailed and accurate modeling of historical sites, leading to a better understanding of their original structure, significance and development over time. Although specific examples of the LIM use in the context of historic landscape reconstruction were not been directly mentioned in the available papers, potential applications and benefits of this approach can be deduced from more general discussions on HLIM and LIM.

The use of landscape information modeling (LIM) in the historical landscape monuments reconstruction in a virtual environment offers several advantages and possibilities:

- **Accurate reconstruction:** LIM allows creating of accurate digital models of the landscape and monuments with a high level of detail. These models can be based on historical data, archival materials, photographic records and archaeological findings, thus representing the past in a more accurate way.
- **Interactive visualizations:** Digital models created using LIM can serve as a basis for interactive visualizations of historical sites and monuments in a virtual environment. These visualizations can be used to present the site history, significance and development.
- **Virtual tours:** Using LIM, it is possible to create virtual tours of historical sites and monuments. These tours allow visitors to explore the past and cultural significance of various sites through interactive virtual environments.
- **Simulations and reconstructions:** LIM allows creating simulations and reconstructions of historical scenes and events. These simulations can help to better understand life and culture in the past and allow visitors to experience history first-hand.
- **Education and research:** Virtual reconstructions of historical landscapes and monuments created by LIM can serve as educational tools and support research in the fields of history, archaeology and cultural heritage.

Reconstructions and visualizations of historic landscapes with monuments indicate enormous potential of LIM. Integration of modern technologies such as LIM in the field of historical research and the cultural heritage protection opens up new possibilities for understanding and preserving our past. The use of LIM and HLIM for virtual reconstruction and experimentation with historical landscape monuments represents a revolutionary approach in the field of archaeology, historical research and the cultural heritage protection. This innovative process not only allows for accurate and detailed reconstruction of historical sites in digital format, but also offers a platform for simulating various environmental processes and for experimenting with the conservation techniques. The LIM-created detailed 3D models provide the basis for virtual restoration of cultural landscapes and monuments, enabling the reconstruction of their original appearance, structure and, last but not least, their context. These models serve as a valuable resource for virtual tours and interactive exhibitions that open the door for the general public to experience and learn history with no need for their physical presence at the site. Simulations of environmental processes, such as climate changes, vegetation changes or erosion, offer unique insights into the factors influencing the historical landscapes development and decline. This knowledge is crucial for scientific research and can help in elaborating the strategies for the monuments sustainable protection and management. Digital experiments with conservation techniques in LIM environment help predict results and identify potential problems before their real-world application. This approach minimizes risks and increases efficiency in the historical monuments protection and conservation. This experimental aspect of LIM represents a significant step forward in the conservation science.

Education and popularization are other consequential aspects of using LIM. Digital reconstructions and simulations offer engaging and interactive ways to educate the public about historic landscape monuments. These tools increase awareness and interest in the cultural heritage protection, which is crucial for its long-term preservation.

Finally, LIM serves as an important tool for planning and development, allowing the potential impacts of new projects on historic landscapes to be assessed. Digital models enable identifying of the most appropriate practices for creating the balance between new development and the cultural heritage protection. Overall, the use of LIM for virtual reconstruction and experiments opens up new avenues for the preservation, study and presentation of cultural landscape monuments. It provides the tools necessary for a deeper understanding of our past and for strategic planning for its protection into the future, underscoring its importance in the present-day scientific and conservation discourse.

Chapter 5

Reconstruction and Experiment

In scientific research, the method of reconstruction is a priceless tool that can find its application in many scientific disciplines. This multidisciplinary approach allows scientists to recover and interpret past events, objects and processes by analyzing available data and material traces that have been preserved to the present. It is aimed in creating the most accurate and complete picture of the past, which provides new possibilities for understanding past contexts, evolution of anthropogenic objects or development of climate change. These methods can be assigned to numerous the so-called productive results of scientific work. They are based on the premise that the project purpose is the basis, while its diversity and relations define its structure, and its solution could be imagery and representation in some form (Benčo 2001, 54).

Importance of reconstruction methods is substantial for both the academic study of the past and also for its applied research, where they can help shape strategies for solving present-day problems and predicting future challenges. In scientific research, these methods are based on a set of theoretical principles that ensure that the process of restoration, modeling or simulation of historical, archaeological, social or environmental conditions and objects is correct in its methodology, verifiable and relevant. They can provide a deep insight into the complex interaction of the science history, technology and material culture (Salvia 2007, 81). It begins from the principle of basic documentation, which emphasizes importance of solid and verifiable data obtained from a variety of sources, including archaeological finds, historical documents, maps and digitization practices. These data are essential for capturing and understanding the original conditions to be reconstructed.

This basis is followed by an interdisciplinary approach, underlining that the reconstruction often requires combined expertise of professionals from various disciplines. The combination of knowledge from history, archaeology, architecture, the IT sector and other disciplines makes it possible to achieve more complex and accurate results. This multidimensionality is connected with the principle of contextualization, according to which it is necessary to place the reconstructed object or event in the correct historical, cultural and social context, so that the reconstruction reflects the reality of past periods.

The scientific process also includes the principle of hypothetical modeling, which allows researchers to fill gaps in the data by using hypotheses supported by science. This approach is essential in situations where direct evidence is lacking and allows for making informed estimations of the past.

At the same time, verification and revision are the key aspects in the reconstruction process, i.e. any result should always be open to updates based on new discoveries or better understanding. This principle supports scientific integrity by ensuring transparency and adaptability of the research process.

Finally, there is an ethical principle that emphasizes the need to carry out reconstructions with respect for the original cultures and their cultural heritage. All procedures should be consistent with the ethical standards and rights associated with reconstructed objects and sites, thus ensuring that the scientific research is both accurate and responsible.

These theoretical principles form a solid methodological framework for reconstruction in scientific research, ensuring that the process is not only academically valid, but also transparent, ethically responsible and open to revision (Diagram 3).

This approach helps the scientific community and the general public to better understand how certain conclusions were reached and promotes confidence in the scientific reconstruction work.

The principles how to carry out the scientific reconstructions can be divided as follows:

- **The principle of induction** is a fundamental building block of the scientific method, which allows generalize observations, as the data are

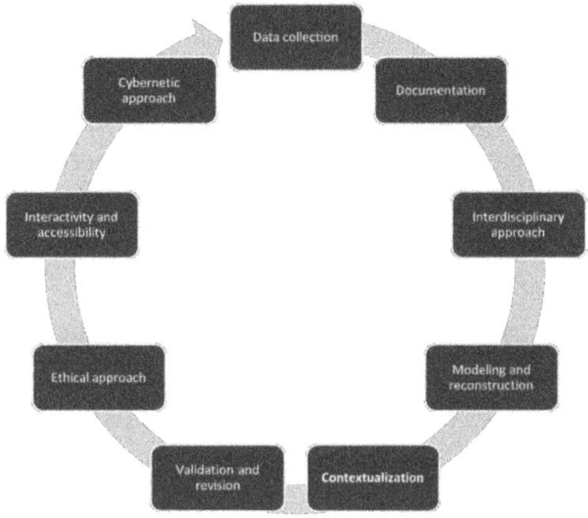

DIAGRAM 3

used to support or refute scientific hypotheses in order to get closer to a true knowledge (Hugh 2012, 267–268). In the reconstruction context, this means to make an effort to draw more general conclusions about past civilizations, ecosystems, or geological processes inductively from a limited amount of data, such as natural facts or artefacts. An essential challenge is to ensure that these inductions are supported by sufficiently solid data and that they are correctly interpreted within the framework of available knowledge.

- **The principle of analogy** consists in the assumption that similar cases have similar causes and consequences. This principle is crucial in archeology and paleontology, where scientists compare finds with modern equivalents to deduce function or environment. For example, eating habits of extinct species can be concluded from the comparison of fossil teeth with those of contemporary animals in their original country.
- **The principle of stratigraphy** is essential for determining relative age of geological layers and archaeological finds. This principle is based on the observation that in sedimentary layers older objects are typically located lower than younger ones. Stratigraphy is the starting point in chronological reconstruction that allows scientists to map historical and prehistoric events in time.

- **The principle of causality** is a fundamental principle that assumes that every effect has its cause. No universal and objective causal criteria, if they exist, have yet been identified (Hugh 2012, 208–209). In the context of reconstruction, this means that each event or phenomenon under reconstruction was caused by specific causes. The challenge is to identify these causes based on available evidence and correctly interpret their influence on the reconstructed phenomenon.
- **The principle of minimizing complexity** (or **Occam's razor**[1]) prefers simpler explanations to more complicated ones, if both can explain the observed data equally well. This principle helps scientists focus on the most likely explanation without unnecessarily adding complexity that is not supported by evidence.

The above mentioned principles of reconstruction in scientific research deepen our understanding of the way scientists approach the reconstruction of past events, objects and processes in a certain geographical space. A completely new and revolutionary move in the reconstruction of a limited geographical space and anthropogenic monuments in it is represented by methods that combine classic methodological procedures with the latest virtual reconstruction technologies. This approach can create accurate and interactive 3D models of past environments that are offering unprecedented opportunities for research, education and presentation of the landscape monuments.

Virtual and digital reconstruction of monuments and landscapes is a multidisciplinary field that uses advanced technologies to create accurate digital models of historical objects and environments. This process is based on various theoretical principles that ensure its scientific accuracy, ethical integrity and practical applicability. Basic theoretical principles that are the foundation of virtual and digital reconstruction of monuments and historical landscapes are analyzed in detail below.

1 Occam's razor is a philosophical and methodological principle that prefers the simplest explanation, which effectively explains the observed facts, to more complex alternatives. This principle is named after the English Franciscan monk and logician William of Ockham (c. 1287–1347), who often used it in his philosophical arguRnts.

1. **Interdisciplinarity**
   - **Integration of disciplines:** Virtual reconstruction requires cooperation of archaeologists, historians, architects, computer scientists and digital art experts. This interdisciplinarity ensures that the reconstruction involves different perspectives and approaches, from historical research to applied technologies.
2. **Authenticity and accuracy**
   - **Basis in primary and secondary sources:** Reconstruction should be based on solid archaeological, historical, ethnographic and environmental evidence to ensure that a digital model is authentic and exact.
   - **Evidence-based modeling:** When creating a virtual reconstruction, available evidence and interpretive models have to be balanced so that the results are credible and at the same time informed of the latest knowledge.
3. **Dynamism and adaptability**
   - **Revision and updating:** Digital models are open to be easily revised and updated in response to new discoveries or better understanding of the matter. This principle supports the dynamism of science, as it allows for the models to be adapted to new scientific findings.
4. **Interactivity and accessibility**
   - **Public involvement:** Digital reconstructions offer unique opportunities for interactivity and education that allow for the general public to better comprehend historical periods and cultural heritage.
   - **Democratization of access:** By using online platforms and virtual reality, information becomes more accessible to the general public, which contributes to spread knowledge and raise appreciation of cultural heritage.
5. **Ethics and sustainability**
   - **Respect for the cultural context:** In the process of reconstruction, it is important to respect the cultural context and origin of historical monuments, and at the same time to ensure that the interpretations are sensitive to the historical and cultural integrity of the objects.
   - **Sustainable digitalization:** In the digitalization process, to think about its sustainability is necessary in the context of long-term storage of digital data as well as in the choice of technologies that minimize its impact on the environment.

The cybernetic approach to virtual heritage and the distinction between physical and virtual reconstruction represent basic concepts that enable the development and application of digital technologies in the field of the cultural heritage protection and conservation. Applying these theoretical principles to the reconstruction of historical landscapes and monuments allows for scientists both to restore the physical appearance of past environments and objects and to understand the social, economic and cultural aspects that produced their development. This approach is based on the idea that a digital object – be it a text message, 2D image, 3D model, film or sound – can transmit information and store knowledge about the object even if its physical original is lost. This approach can save the cultural heritage by digital archiving and documenting, and simultaneously it opens up new possibilities for its presentation and interpretation. In the context of cybernetics, a digital object is considered not just a static replica of a physical object, but rather part of a dynamic space of relations and interactions. This approach changes traditional notions of objectivity and raises new questions about the materialization, presence and reproduction of scientific objects (Salvia 2007, 81). The virtually reconstructed model is thus not just a passive representation of the past, but an active environment that stimulates understanding and attribution of significance through interactive experiences. Therefore, the aim is both to visualize the past and to create a space for the user's active participation stimulating his critical thinking and creativity. The differences can be also found between physical and virtual restoration and reconstruction, which is another essential aspect in the process of preserving the cultural heritage. Physical restoration refers to methodological interventions into an art work material with the aim of preserving or restoring it, while virtual restoration uses digital techniques for the rescue, analysis or reconstruction of objects and structures without physical intervention in their material essence (Pietroni/Ferdani 2021, 2–3).

Thus, virtual reconstruction is a process in which digital models are used to visually restore objects or structures from the past based on existing physical evidence and scientifically justified inferences. This approach allows the presentation and interpretation of cultural heritage in a new light and at the same time provides tools for scientific research and education, enabling deeper comprehension of historical contexts and technologies used in the past. Virtual reconstruction of historical landscapes and monuments includes diverse approaches and technologies that are used to capture, model and visualize historical objects and environments. Virtual reconstructions are mostly based on 3D laser scanning or photogrammetry techniques, detailed in the previous chapter, which produce geometrically accurate photorealistic visualizations of objects and space. Both approaches – the cybernetic approach and the distinction between physical and virtual restoration and reconstruction – thus represent two sides of the same coin in the process of studying, preserving and presenting cultural heritage. Together, they open new opportunities for research, education and access to historical and cultural values for the general public, respecting the needs of protection and sustainable use of cultural heritage in the digital age (Pietroni/Ferdani 2021, 3–4).

Although the reconstruction methods render valuable insights into the past, they are also associated with various challenges. The substantial ones are those of interpretation of incomplete or damaged data, risk of anachronism and need for multidisciplinary approach to data acquisition and analysis. The data interpretation demands thorough understanding of current scientific knowledge as well as creativity and ability to see interconnections where they are not obvious at first sight. Moreover, the scientific community has constantly to face up to prejudices and expectations, which can influence the interpretation of evidence. The method of reconstruction is an integral part of scientific research, providing unique perspectives on our past that can help us better understand the present and prepare for the future. Despite the challenges and limitations the researchers have to face, continued advances in technology and methodology open up new possibilities for more accurate and detailed reconstructions. Thanks to these advances, our understanding of the past is expected to continue improving, bringing new perspectives to various aspects of historical development and future prospects.

Experimentation as a scientific method is another key pillar of scientific research that also represents an integral part of the knowledge development in various professional disciplines. Its application provides scientists with possibility to test hypotheses and theories systematically by means of precise observation and experimentation, thus ensuring objective, repeatable and verifiable results. This complex process is based on a series of methodological steps that are essential for empirical verification of scientific assumptions (Benčo 2001, 19). Experiments can serve to articulate a paradigm, to explore and extend its application to new areas, or to detect anomalies that may prompt the need for the paradigm to be revaluated or changed. By this way, a theory and experimental practice can be deeper connected in the scientific research, where experiments are not only tools for theories to be tested, but they are modeled according to theoretical assumptions and expectations (Kuhn 1970, 23¬26).

Each scientific process begins with observation. Intrigued by curiosity, scientists use to investigate natural phenomena, technological processes, or social interactions, looking for unusual, unexplained, or unexplored aspects. Scientific knowledge is largely, but not entirely, dependent on observation, experimental evidence, rational arguments, and skepticism (Hugh 2012, 234–236). The observation raises questions and leads to identification of a specific problem that requires further investigation. This moment of curiosity and questioning is the cornerstone of scientific discovery. Once the problem is clearly identified, it is followed by formulating a hypothesis. Basically, the hypothesis is a testable assumption that offers a possible explanation for the observed phenomenon. It takes the form of a statement that suggests a relation between independent and dependent variables. The competition between two hypotheses is discussed: the currently preferred hypothesis and the newly modified one. It discusses how the logic of deduction and induction can be used to compare different hypotheses and determine which one is true or likely to be true considering the data obtained (Benčo 2001, 31–32; Hugh 2012, 267–268). The scientific rigor requires that a hypothesis has to be formulated in such a way that it can be unequivocally verified or disproved through experimental tests.

The experiment is the heart of a scientific method. In this phase, researchers design an experiment that will allow them to test their hypothesis under controlled conditions, where selection and control of variables is important. The researchers have to ensure that any changes in dependent variable are the result of manipulation with independent variable, while all other variables have to be constant or controlled. The thorough planning and preparation is followed by actual experiment implementation. The current popular hypothesis influences the design of a relevant experiment to generate empirical observations. In designing an experiment, the scientist has to consider the prospective value of his data to discriminate between competing hypotheses, along with the experiment costs and risks. Well designed experiments minimize the negative effects of experimental errors

(Hugh 2012, 267–268). Experiments can be conducted under laboratory or field conditions and often involve repeating to guarantee accuracy and reproducibility. In this phase, the data collection is critical as it provides the empirical basis for evaluating the hypothesis.

The collected data are subsequently subject to thorough analysis. There are various statistical methods used to process and interpret data, which allows determining whether, according to the experiment results, the original hypothesis can be proved or disproved. The analysis may also reveal unexpected patterns or connections that can lead to new questions and hypotheses. Based on the data analysis, the conclusions are formulated that summarize whether the hypothesis has been confirmed or refuted. This phase also includes critical reflection on the experiment, identification of possible limitations of the study, and suggestions for further research. Publishing the results in professional journals is the next critical step. This process includes review by experts who independently evaluate the research, its methodology and conclusions. The peer review ensures that published results are reliable and beneficial to the scientific community.

The science values repeatability. Repeating the experiments by other teams or in different contexts helps to validate the results and guarantee their objectivity and reliability. This aspect of the scientific process underlines collective nature of scientific knowledge, where each new research contributes to a wider understanding and confirms or refutes previous findings. The scientific method of experimentation is a dynamic and iterative process that allows the scientific community to continually build on the foundations of previous knowledge. Its systematic approach ensures that scientific findings are evidence-based and can be continuously tested and extended in the context of never-ending search for truth. For example, S. Salvia (2007, 97) points out the importance of selecting the materials, implementation techniques and tools that may or may not be used, while all aspects related to technical implementation in this case have to be processed with the highest possible degree of philological accuracy that the state of our current historical knowledge allows. Digitization and technological advances provide the scientists with tools and platforms that allow explore, test and analyze hypotheses with unprecedented precision and efficiency. In this comprehensive and wide-ranging text, we explore how the digital space is transforming the paradigm of scientific experimentation, expanding its possibilities, and presenting the challenges for scientists to cope with.

Traditional scientific experiments, realized in laboratories with physical instruments and materials, remain the foundation of science. However, the entry of digital technologies into the field of scientific research creates a new dimension of experimentation. Computer-based simulations, high-performance computing, and artificial intelligence (AI) algorithms allow modeling of complex systems and processes that would be physically demanding, time-consuming, or costly to investigate in the real world.

Digital experiments can simulate various conditions and situations, allowing scientists test their hypotheses in virtual environments where variables can be precisely adjusted and manipulated. This gives the opportunity for more accurate and broader testing of theories and can lead to discoveries that would otherwise be beyond the reach of traditional methodologies.

Thanks to the development of digital tools, scientists can now conduct experiments that were previously unimaginable. Virtual reality (VR) and augmented reality (AR) offer new ways of interacting with experimental data and models, allowing scientists to "enter" their experiments and observe them from new perspectives. The Big Data and data analytics provide access to vast amounts of information that can be analyzed to reveal patterns and connections that would otherwise remain hidden.

Automation and robotics are also changing the space of experimental science, enabling automated experiments that can run continuously or with minimal human intervention. This not only increases the efficiency and accuracy of experimental procedures, but also opens the door to experiments in environments that are too dangerous or inaccessible for humans. However, the advancement in digital experimentation brings some challenges. Data protection, privacy and ethical issues are becoming increasingly urgent, especially when experiments require collecting and analysing of large amounts of personal or sensitive data. Scientists and institutions have to be vigilant in protecting this information and ensuring that its use complies with ethical standards and legal regulations.

In addition to this, the dependence on digital tools and platforms raises questions about the reproducibility and verifiability of the results. To maintain the integrity of scientific research, it is crucial to ensure that digital experiments are transparent and their results can be independently verified.

The possibilities of scientific experimentation in the digital space represent a revolution in scientific research. They push the boundaries of what can be explored and comprehended, opening new avenues for discoveries that can transform our understanding of the world and the universe. However, new challenges come with these possibilities that require attention, caution and ethical consideration. In this digital era of scientific experimentation, we have to be prepared to face these challenges while fully exploiting its enormous potential to advance knowledge.

Chapter 6

Time dimension and historical context

In virtual reconstruction, dimension of time is a key aspect in the process of creating historical monuments, landscapes or events through using digital tools and technologies. This concept focuses on the ability to accurately depict and simulate how specific sites, objects or monuments have changed and developed during different historical periods. The dimension of time adds depth and context to virtual reconstructions, allowing not only to see but also to understand the dynamics of change in the course of time. This includes, for example, showing how architectural styles evolved, how natural disasters or human interventions affected the landscape, or what were the impacts of technological advances on social structures and people's daily lives.

The importance of historical context and its function in the process of reconstruction of monuments and historical landscapes is a fundamental mainstay for any effort to restore the past authentically and accurately. Historical context encompasses the set of all circumstances that influenced specific historical events, objects or personalities, including political, social, economic, cultural and geographical factors. These factors have profoundly determined the course of history and are necessary for correct comprehension and interpretation of past periods (Diagram 4).

Time dimension

Correct understanding of historical context is essential for both the scientific accuracy and preserving the credibility and authenticity of reconstructed

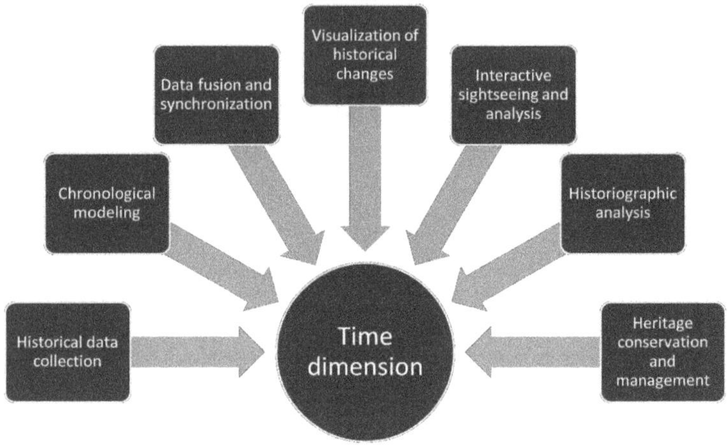

DIAGRAM 4

monuments and landscape units. Without deep knowledge of historical context, reconstructed projects would be superficial and potentially misleading, which would have a negative impact on education, scientific research and monuments as well.

Appropriate selection of methodological approaches is crucial to achieving a veracious reconstruction. Archival research, which involves working with historical documents, deeds, maps, photographs and other written sources, is the basis for understanding the written testimonies of the past. Critical analysis and interpretation of these materials requires advanced knowledge of historical methodologies and the ability to analyze information in a broader context. Archaeological or monumental reconstruction adds another layer of knowledge by providing physical evidence of the past that can complement or even challenge the records found in archival documents. The use of techniques such as stratigraphy and analysis of material cultures allows for a more accurate location of historical objects and events in time and space.

Introducing the fourth time dimension into the three-dimensional geometric modeling of real data enables creation of a multi-temporal representation of the site. In this way, a new set of tools and working methods is provided for the study of monuments development, advancement of hypotheses about the past and modeling of probable future course. The ability to see dynamic development of landscape-historical anthropogenic objects at various spatial scales (e.g. building, locality, city or territory) summarized in a diachronic model enables better comprehension of the monument current state with respect to its chronological development (Rodríguez-Gonzálvez et al. 2017, 609; Rodríguez-Gonzálvez et al. 2018, 119; Głowienka 2017, 1–5).

Multi-temporal 3D reconstruction is a process that creates a series of three-dimensional models of cultural heritage (CH) or monuments that capture and represent their condition in different historical periods. This approach allows scientists, historians, architects and other professionals to analyze and understand the development and changes that occurred in monuments in the course of time. Substantial aspects of multi-temporal 3D reconstruction according to (Rodríguez-Gonzálvez et al. 2017, 610–611) are following:

1. **Data collection:** The first step is collection of historical and current data about a given landscape-historical or architectural object. These data can be obtained from diverse sources, including old maps, photographs, drawings, written records and contemporary 3D scans.
2. **Data fusion:** Subsequently, data from different sources and time periods are combined and integrated into a unified model. This process may demand advanced techniques to reconcile and calibrate data from different scales and perspectives.
3. **Modeling:** Using specialized modeling software, the obtained data are transformed into detailed 3D models that accurately reflect the appearance and structure of digital objects in different historical periods. This step can include modeling losses, restoring missing parts and simulating historical changes.
4. **Visualization and analysis:** Created multi-temporal 3D models can then be visualized and analyzed to gain new insights into historical development, construction techniques, change of use and impact of environmental factors on the given objects under research.

5. **Interactivity and sharing:** Modern technologies such as virtual and augmented reality (VR/AR) allow interactive viewing and exploration of 3D models. These models can be shared with the academic community, students and the general public, increasing awareness and education about cultural heritage.
6. **Heritage conservation:** In addition to research and educational applications, multi-temporal 3D reconstructions provide a valuable opportunity for conservators and restorers in the process of planning the monuments conservation and maintenance.

In addition to traditional archival and archaeological methods of data collection, the analysis of historical sources also includes the study of art, literature, folklore and other cultural manifestations that can reflect the social values and norms of a specific historical period (El-Hakim et al. 2006, 1–2). This multidisciplinary approach brings a more complex and deeper understanding of the historical context, which is essential for creating veracious and objective models of the past. Therefore, one of the key challenges in 4D reconstruction and visualization of cultural heritage is integrating the multi-source data. This process requires efficient combining of different types of data, including metric and non-metric ones, to create dynamic representations that capture changes in monuments and historic sites over the time. In order to achieve effective integration, various data sources have to be approached systematically, considering the size and complexity of a particular object or structure (Rodríguez-Gonzálvez et al. 2018, 120–121).

The importance of the time dimension in virtual reconstructions is justified from several methodological aspects:

- **Historical accuracy and contextualization:** Integration of the time dimension into the process of virtual reconstruction represents a revolutionary step in our understanding and interpretation of history. This methodology gives a possibility both to date objects and monuments relatively accurately and to display their gradual development in the course of time. The sequential view provided in this way contributes to a deeper and more complex historical context, revealing layers of information that would otherwise remain hidden. The time dimension allows users and researchers not only to see, but also to understand the dynamics of change that took place at specific sites or with specific objects. This insight into history becomes even more relevant when the development of architecture, technology and social structures during different historical periods is examined (Rodríguez-Gonzálvez et al. 2018, 119). The virtual view makes it possible to analyze how building styles changed, how new technological innovations were introduced, and how social arrangements and relations were transformed. Tracing these changes in specific historical contexts not only broadens our knowledge of the past, but also contributes to a better understanding of the processes that have shaped contemporary social and cultural structures. It allows us to reconstruct and visualize the past in its full complexity, and at the same time discover the connections and influences that shape our present.

- **Interactivity and engagement:** Virtual reconstructions with added time dimension bring new level of interactivity and engagement to the cultural heritage and education segment. These reconstructions, enriched with interactive elements, provide users with a unique opportunity to explore different historical periods and observe the changes that took place in specific sites completely independently. This personal approach allows users not only to see, but also to "experience" history as if they were moving through it themselves. The possibility to "travel in time" and see the transformation of sites or objects in virtual reality allows deeper understanding of historical contexts and processes (El-Hakim et al. 2006, 1). This interactive component considerably supports a learner´s or visitor´s interest and engagement. Having the opportunity to participate in their own learning process actively, people are more interested and motivated to learn more. The interactivity of virtual reconstructions transforms passive observation into active discovery, which makes the learning process more attractive and effective at the same time. Furthermore, the possibility to experiment with historical scenarios, change perspectives or even engage in virtual simulations of historical events adds to the educational value and helps to consolidate the acquired knowledge.

The expansion in this concept can also include the use of game elements such as challenges, tasks and quizzes, which are put into historical contexts, allowing users to test their knowledge and skills in practice. Thanks to recent advancement in the gaming industry that have brought affordable VR and AR devices, these technologies are becoming increasingly popular outside of video games, too. VR is proving useful for the representing the cultural heritage in time and is beginning to play an important role in museums in simulating different historical scenarios, thus providing the possibility to re-examine them (Rodríguez-Gonzálvez et al. 2018, 136). Such approaches can further increase interest and engagement by transforming learning into an interactive game where each new piece of information is part of a larger story or adventure.

- **Visualization and simulation of changes:** The method allows detailed demonstrating of successive transformation of a specific site in the course of time, as a result of both the natural events and human activity and technological progress. In this way, we can observe how cities and landscapes developed, how architectural styles changed, or how landscapes were transformed by human activity and interventions into natural environment. This view allows better understanding the past and at the same time pondering the current relationship of man to nature (El-Hakim et al. 2006, 1).

  In addition to the visualization of changes, the time dimension in reconstructions also enables simulating of historical events. In this way, we can virtually experience important historical moments, such as battles, important construction projects, or key social changes. Simulating these events offers the opportunity to gain a deeper comprehension of their impact on society, culture and physical environment (Champion 2021, 129–139). For example, simulation of the historical monuments construction can reveal not only the technological innovations used in the given time, but also the socio-economic conditions that enabled or influenced their construction.
- **Education:** In pedagogical practice, the virtual time dimension provides a dynamic and visually appealing way to learn about the past, greatly widening the possibilities of understanding and exploring historical events, trends and processes. It is not just a static depiction of past periods, but a dynamic story manifesting how and why changes occurred during the time. This ability to visualize historical development provides deep insight into the ways how societies, technologies, architecture and landscapes gradually developed and changed. By expanding virtual reconstructions by the time dimension, new possibilities are opened for teachers and students. The teachers can use these tools to create interactive lessons that allow students to not only see, but also "relive" historical periods and events. In this way, the learning process becomes more interesting and engaging, which significantly increases the knowledge retention and understanding.

In addition to educational benefits, virtual reconstructions with a time dimension can also offer significant advantages for research. Scientists can use these tools to test hypotheses about how various circumstances may have influenced the development of specific sites or objects. They can simulate different scenarios to observe how certain changes could affect the appearance and functionality of cities or buildings. This approach can reveal new information about historical processes and provide new insight into the connections between the past and the present. Moreover, these technologies can serve as a bridge between academicians and the general public, increasing public awareness and appreciation of cultural and historical heritage. Virtual reconstructions with a time dimension allow people all over the world to discover and learn about the history in a way that was previously unimaginable.

- **Reviving the past:** This approach opens the door to comprehending how historical events, decisions and trends are reflected in our contemporary world. It allows us not only to see changes in landscapes, architecture or social structures, but also to realize the processes that led to the current situation. In this way, the temporal dimension provides a valuable context that helps better understand current challenges and trends in the light of their historical roots.

In addition to providing context and highlighting the connections between the past and the present, visualizing history in time has a strong emotional impact as well. When users are able to watch and "live" historical moments through virtual reconstructions, it brings people's stories and events closer to them on a much more personal level. This emotional experience can arouse a deep interest in history, promote empathy and understanding for past generations, and show the impact the historical decisions had on people's lives.

Expanding of this concept offers opportunities to create educational programs and exhibitions that convey historical knowledge not only through facts and data, but also via emotional and interactive experiences. The history revived in this way can inspire a new generation of historians, researchers, educators and the general public to actively engage in the cultural heritage protection and conservation. In addition, the emotional impact and visualization of interconnections between the past and the present can help build a collective identity and strengthen a sense of belonging to a particular place or community.

- **Support for monuments conservation and restoration:** In the process of cultural monuments restoration and conservation, virtual models can be used by professionals to predict potential outcomes of their interventions, what is increasing the accuracy and efficiency of planned processes. With the help of time dimension, restorers and conservators can virtually simulate different approaches and intervention techniques and thus optimize the decisions regarding the monuments protection and conservation with regard to their historical value and integrity (Głowienka 2017).

  In addition to this, the dynamic display of time development provides a valuable tool for effective documentation of changes in the monuments and landscape conditions. This process makes it possible to capture and visualize the gradual changes that monuments undergo, whether as a result of natural aging, environmental influences, and human activity or restoring interventions. The information obtained in this way is invaluable for historical research and provides an important basis for strategies aimed in the protection and sustainable management of cultural heritage.

  Expanding use of time dimension in reconstructions supports transparency and public engagement in the process of monument protection. Presenting the planned restoration works and possible results through virtual models enables the lay public to better understand the value of historical monuments and the need for their protection. This can lead to wider public support and interest in the cultural heritage, and better understanding of the challenges associated with its preservation as well.

  Finally, the virtual reconstructions provide a platform for international collaboration of experts in restoration and conservation practice and historical research, enabling the sharing of knowledge, techniques and tried methods. This promotes the exchange of expertise and cooperation on a global level, which is crucial for the cultural heritage protection and conservation for future generations.
- **New perspectives in interpretation:** The possibility of exploring various time periods and scenarios allows both the historians and the general public to gain insight into alternative interpretations of history, thus expanding the traditional understanding of historical events. This approach gives the users opportunity to see how certain historical moments might be different under changed circumstances, providing a unique opportunity to

gain a deeper understanding of the past events dynamics and complexity. This opens up new perspectives for comprehending the past, encourages critical thinking and stimulates curiosity and discussion about how history and its various interpretations affect our lives today.

The time dimension in reconstructions also supports interdisciplinary approaches by stimulating collaboration between experts from different fields, such as history, archaeology, geography, architecture and others. This cooperation leads to a wider and more multi-layered understanding of historical sites and events. When different expertise is brought together, it can lead to new discoveries and interpretations that might otherwise be overlooked within a single discipline. Interdisciplinary approaches allow a better understanding of the context, in which events took place, and the factors that influenced them, thus providing a more comprehensive picture of the past. This capability offers a new set of tools and working methods to support the study of the evolution of cultural heritage sites, to develop hypotheses about the past, and to model presumable future development Rodríguez-Gonzálvez et al. 2018, 140).

Applying the time dimension in virtual reconstructions opens promising possibilities for deepening our understanding of the past and the cultural heritage. However, this approach faces a number of challenges, ranging from technical and financial obstacles to methodological dilemmas and ethical issues. Overcoming these obstacles will require considerable effort and innovative solutions.

One of the main challenges is securing the necessary technical and financial resources to create veracious and detailed virtual reconstructions. High costs of advanced software and hardware, as well as the need for specialized expertise, can limit access to these tools. Furthermore, keeping these models up-to-date and compatible with the latest technologies is a permanent challenge that requires continuous investments in updates and maintenance.

As a result of advances in digital technologies and increasing use of virtual and augmented reality, data optimization is becoming essential in creating smooth, interactive and visually engaging 4D applications. The optimization begins at the metadata level, where it focuses on simplifying and structuring data information, making it easier to manage and access. In the case of digitized and 2D digital data, to reduce the resolution and optimize the database structure in order to be able to store and search the data more efficiently is necessary. In order to achieve effective management, visualization and interactive presentation of cultural heritage in 4D, thorough work on data optimization is essential. The practices such as model segmentation, use of normal and displacement maps, and UV unfolding techniques are substantial for both the achieving effective visualization and ensuring a smooth experience for the user (Rodríguez-Gonzálvez et al. 2017, 613–614).

Rigorous methodology and historical credibility are another challenge. It is crucial that virtual re-enactments are reflecting historical events accurately and are backed by thorough research. This process has to consider multifarious historical interpretations and balance them in order to avoid one-sidedness and to present balanced interpretation of the past. Increasing accessibility and promoting public participation are other key areas where efforts need to be focused. It is essential that virtual reconstructions are accessible to a wide range of people, including those with specific needs, and that active public involvement in the process of creating and revising virtual models is encouraged. It is necessary to identify the specifications, needs and demands of the landscape monuments in order to understand the required levels of information of the 4D model, thus allowing to determine optimal materials and technologies to be used at different spatial scales, and to set the data management and visualization requirements (Rodríguez- Gonzálvez et al. 2017).

In order to overcome these challenges, technological innovation and increased investment are required along with multidisciplinary cooperation of historians, archaeologists, technologists and designers. The potential of time dimension in virtual reconstructions can be maximized by working together and focusing on sustainability, inclusivity and ethical responsibility, by which way new possibilities in the research and presentation of monuments and historical landscapes can be open.

Chapter 7

Selected methods of virtual reconstruction of historical landscape and monuments

Virtual reconstruction uses wide range of digital tools – from 3D modeling and animation to augmented and virtual reality. These technologies offers to historical geographers, conservationists, historians and archaeologists to reconstruct destroyed or partially preserved historical objects, structures and landscape units with distinguished exactness and detail. Interactive experiences as a part of virtual reconstructions can arouse interest and curiosity in people of all ages, thereby democratizing the access to historical and cultural heritage. Despite the indisputable benefits, virtual reconstruction also encounters some challenges. Questions regarding the reconstruction exactness and authenticity, the ethical aspects of using the historical records and technological limitations are the issue of professional dispute. Regarding the virtual reconstructions, strict scientific procedures have to be applied and an interdisciplinary approach followed to ensure the greatest possible degree of historical relevance and scientific rigour (Schulze-Horsel 2007, 1).

In this assignment, digital hermeneutics is one of the crucial concepts (Romele/Severo/Furia 2018, 73–75), which applies principles of hermeneutics, i.e. art and theory of interpretation, to digital environment. Digital hermeneutics is coalescence of hermeneutics and web technologies, where the traditional theory of interpretation in the humanities is adapted to digital environment. This approach is important for accessing and interpreting online cultural heritage collections (Akker et al. 2011, 1; Evink 2020). It also provides a new way of analyzing and interpreting historical data by using digital tools for a better understanding of cultural heritage. Thus, digital hermeneutics is connecting traditional hermeneutic techniques and the potential that digital technologies can offer to deeper understanding of historical artefacts (Romele/Severo/Furia 2018, 75–77; Akker et al. 2011). Digital media and technologies are significantly influencing the social reality and the way we interpret and interact with the environment (Evink 2020, 8–9).

Virtual archaeology (Reilly 1990, 133; Forte 2007; Karlsson 2013; Mara 2022), another fundamental concept, uses digital tools and methods to simulate and reconstruct archaeological sites and objects. Methods such as 3D scanning and photogrammetry are cardinal to precise recording and reconstructing the physical characteristics of archaeological finds and monuments. These techniques provide the basis for creating digital twins that can then be analyzed and presented without the risk of damaging the original artefacts. This process not only protects physical monuments from degradation, but also allows the scientific community and the general public to access valuable historical data without geographical or time restrictions.

The authenticity and interpretation are critical implications that must be considered in virtual reconstruction. While digital technologies offer unique possibilities for reviving the past, it is essential to maintain scientific exactness and integrity in the creation of these reconstructions. Ethical issues, such as correct representation of historical facts and cultural contexts, are integral part of the digital reconstruction process, while the researchers have to balance the technology used to bring the past to life with a responsible way of interpreting the history.

In the field of landscape architectural design, visual research is a key implement for the landscape development and interpretation. Modern technologies, such as geographic information systems (GIS) and computer-aided design (CAD), open up new possibilities for visualization, analysis and interpretation of landscape structures and processes. These tools enable veracious modeling and simulation of landscape interventions at various scales, which is essential for identifying design options and assessing potential impacts on the landscape environment (Nijhuis et al. 2011, 103–120).

Visual research in landscape architectural design is invaluable for identifying and solving specific problems in the landscape, such as visual barriers, perspectives and integration of new elements into the existing landscape context. The importance of a multidisciplinary approach that integrates ecological, social and cultural aspects with aesthetic and design goals is an inherent part of this process. This approach gives the opportunity of creating sustainable and aesthetically appealing landscape solutions that respect and enrich the already existing cultural and natural values (Nijhuis et al. 2011, 121–134).

The challenges associated with visual research include the need of balancing the technical precision with interpretive flexibility. Involving the community and project participants into the process of planning and designing is essential to ensure that the resulting solutions will reflect all needs and expectations. This aspect emphasizes the importance of a participatory approach in the landscape designing process.

New perspectives in the field of visual research are opened by advanced technologies, such as virtual reality, augmented reality and mixed reality (VR, AR and MR), which can offer new possibilities for the landscape design presentation and exploration (Benvenuto et al. 2024, 3–5). These technologies allow interactively explore various design options and better understand the possible impacts of interventions on the landscape. The integration of VR, AR and MR into the visual research process brings a deeper understanding and better interpretation of the landscape projects (Nijhuis et al. 2011, 135–146; Benvenuto et al. 2024, 12–13; Interrante/Höllerer/ Lécuyer 2018; Fitria Nur 2023).

Concerning the interpretation of published sources, the visual research playing a fundamental role in the process of landscape architectural design is obvious. Using of modern technologies enables the landscape visualizing and analyzing to be more effective, thus opening new ways for its development and interpretation. Participation of the community and multidisciplinary approach are essential to creating landscape solutions that are sustainable, aesthetically valuable and reflecting the needs and expectations of the general public. Creating three-dimensional (3D) models and photorealistic visualizations of the environment is relevant in many applications, starting with virtual landscape representation to land cover/use analysis and monitoring, as well as landscape management and planning (Dinkov/Vatseva 2016, 320). Development of new technologies, such as VR and AR, provides implements for better comprehension and presentation of the landscape projects, which contributes to their better interpretation and understanding (Benvenuto et al. 2024, 3–5).

The technologies such as 3D geographic information systems (GIS), computer-aided design (CAD) and advanced computer graphics are essential part of the reconstruction process. These tools enable creating of exact and detailed models of historical landscapes that can serve basis for the cultural heritage research, education and presentation. Three-dimensional GIS technologies allow integrating of various types of data, including topographic maps, historical texts and archaeological findings thus creating a comprehensive picture of past landscape structures. CAD and advanced computer graphics contribute to the accuracy and aesthetic value of the reconstructions, which can increase their potential as a source of education and presentation (Sadrehaghighi 2022).

The main challenges of virtual reconstructions include the lack of historical data and need to balance photorealism and abstract representation. The lack of historical data can limit the exactness and completeness of virtual reconstructions, which demands innovative approaches to the interpretation of available sources and to integration of different types of data. The balance between photorealism and abstract representation is crucial for achieving a realistic perception of virtual environments. While photorealism can provide users with an engaging and compelling experience, abstract depictions can be useful in highlighting specific historical or cultural aspects of the landscape (Nijhuis et al. 2011, 186–202).

In the virtual reconstruction of historical landscapes and monuments, numerous advanced technologies and methods are applied, which together form a methodological framework for exact and interactive representation of the past. Substantial techniques include 3D modeling, virtual and augmented reality (VR/AR), geographic information systems (GIS) and other digital tools. Each of these methods comes with its own unique potential and limitations that have to be considered when applying them. The 3D mapping involves four general steps: concept, modeling, symbology and visualization, which affect final quality of the map and proceed iteratively. In the conceptual phase, the user context and product specifications need to be defined. The choice of map type depends on the knowledge and experience of users, possible future users and the necessary geospatial data. The modeling involves processing the original thematic data sets (both raster and vector data), including transformations of data format and structure, object geometry and semantics. Based on the original geospatial data, a complex structured 3D model of the landscape is created. Symbology involves adding graphic attributes to individual objects of the 3D landscape model. In this way, the graphic appearance of each class of objects is defined, like in a classical maps legend, while cartographic rules of generalization are considered. In the last visualization step, the cartographic 3D model is displayed in perspective on the specified media. Cameras are used to set the view position or moving view path. The final rendering process transforms the cartographic 3D model into 3D map with projection parameters and simulation of lighting and shading effects to obtain a photorealistic image (Dinkov/Vatseva 2016, 321–322).

Despite the large number of options to process data obtained by field or cabinet research in digital environment, some selected procedures can be possibly used in a very effective way. Each of them assumes the use of several computer programs and several types of digital data, represented by different formats. However, the resulting reconstruction will be relatively the same in any case, or the results of digital experiments will be slightly different, provided the procedures are used correctly. Proper integration of diverse data from sources is essential. For example, combining LiDAR data, which provides a detailed 3D terrain model, with historical maps and photogrammetric data allows creating of a historical site comprehensive picture. The next important step is to apply specialized software adapted for specific tasks. For example, 3D modeling software such as Blender or AutoCAD can be used to create detailed models of buildings or artefacts based on photogrammetric data (Schulze-Horsel 2007, 2). On the other hand, data processing software such as ArcGIS or QGIS is invaluable for manipulating geospatial data and creating detailed maps of historical landscapes. The goal of 3D landscape modeling is to produce sufficiently reliable and precise digital three-dimensional model of the earth's surface according to the user's needs, which can be dynamically visualized through selected software or can be used to create static cartographic or presentation products (Dinkov/Vatseva 2016, 320).

In this monograph, selected methodological procedures leading to the relevant reconstruction of spatial data and objects in virtual space will be presented in detail, each of them explained on individual model cases.

7.1 GIS (Geographic Information Systems)

Regarding the issue discussed, systems working with spatial data are substantial. Geographic information systems (GIS) represent an important tool in many fields dealing with monuments and cultural landscapes. Due to the system´s ability to integrate, analyze and visualize geospatial data (e.g. satellite images, LiDAR, aerial photogrammetry, etc.), GIS opens up new possibilities for pinpoint data analysis and synthesis (Gáspár 2016; Benvenuto et al. 2024, 20). Therefore, GIS technologies are priceless in the spatial data analysis and visualization, which allows generating of new information based on existing data. GIS provides analysis and spatial modeling tools that are crucial for the historical monuments protection (Rząsa/ Ogryzek/ Ciski 2019, 1).

The first step in using GIS in this context is historical and archaeological data collection and analysis, which may include maps, textual documents, photographs, satellite images and field survey results. This data are subsequently digitized and integrated into GIS environment, where it can be geospatially aligned and analyzed. This process allows identifying and reconstructing of historic structures, roads, waterways, and other cultural landscape features with anthropogenic structures (Haller 2010, 166). One of relevant advantages of using GIS is the ability to layer different types of data, allowing for a complex analysis of historical landscape structures. For example, archaeological contexts can be combined with historical maps and present-day terrain data to obtain a multidimensional view of a given site. This integration of data can reveal relations and dynamics that would otherwise remain hidden, helping to better comprehend the historical processes that shaped the landscape.

Furthermore, GIS is indispensable to modeling landscape changes in the course of time. With its ability to manipulate time series data, the changes can be visualized that took place at specific sites due to natural processes or human activity. This quality is crucial for historical landscape reconstruction, as it allows for both displaying the past development stages of the landscape and simulating future changes or the impacts of various conservation and development strategies. It provides a platform for creating interactive and immersive presentations of historical landscapes and monuments (Flamary et al. 2015, 1). By integrating GIS data with virtual and augmented reality technologies, detailed 3D models of historical sites can be created that users can virtually walk through and explore from every point of view. These models can be further complemented with multimedia content, such as texts, photos or videos, providing a comprehensive experience of visiting the historical sites and learning about them.

For the needs of virtual 3D modeling of historical landscapes and monuments, various methods and data obtained by means of geographic information systems (GIS) can be used. These methods and data provide relevant information required for creating veracious and realistic 3D models. In this work, some methodological procedures were selected to fulfill our needs aimed in the reconstruction of historical landscape environment.

7.1.1 Geospatial analyzes

Geospatial analyzes represent an extensive field of research that deals with the study of objects, phenomena and their mutual relations in geographic space by using geospatial data (Vaughn 2019, 1). Professionals in various disciplines, including geography, cartography, geoinformatics, and remote sensing, rely on these analyzes in the process of interpreting, understanding, visualizing, and predicting phenomena and relations in a geographic context. The starting point in their work is geospatial data, which are bearers of information about location and attributes of objects or phenomena on Earth (Rzasa/Ciski/Ogryzek 2019, 61). The geospatial analysis includes a variety of methods that can be divided into six categories: queries, measurements, transformations, descriptive summaries, optimization, and hypothesis testing. These methods make it possible to obtain information from databases, perform numerical measurements of geographic entities and their relations, change and compare datasets, apply statistical methods to geographic data, optimize the location of points according to defined rules, and generalize from a small sample to the entire dataset (Haller 2010, 167–169). These data can be categorized as vector data, raster data, or multidimensional data, depending on the type and nature of the information they contain (Haller 2010, 169–170). The basic selected methodologies that are used include:

- **Spatial analysis** in GIS is a crucial method for examining and interpreting data in the context of their geographic location and relations. This technique allows users to analyze geospatial patterns, trends and dynamics at various scales and applications, from environmental monitoring to urban planning (Hao 2019, 2–6). It is based on the ability to identify and quantify relations among geospatial objects. For example, the analysis may find out how probable a flooding is in a certain area based on previous precipitation data and terrain topography. Another example is fixing of the optimal route for new transport infrastructure with respect to existing roads and population centres (Rzasa/Ciski/Ogryzek 2019, 61–62). Spatial analysis uses various techniques and tools, which include, for example:

- Buffer analysis
- Overlay analysis
- Network analysis

Despite its usefulness, it is important to emphasize that spatial analysis can be limited by the availability and quality of input data. Poor quality or inaccurate data can lead to erroneous conclusions, which can result in inadequate decisions. Therefore, careful preparation and validation of all data entering the analysis is critical. In recent years, there has been a rapid increase in the use of spatial analysis in various fields, including national and local government, law enforcement, public health, natural resources management, and social and economic analyses (Hao 2019, 2–6).

- **Geostatistical analysis** is a relevant method in the field of geographic information systems (GIS), which is specialized in the study and interpretation of spatial data with regard to their geographic location and mutual relations. This approach is also extremely useful in the context of monuments in cultural landscape, where it enables better understanding of distribution and dynamics of cultural artefacts and their interaction with surrounding environment. One of the basic concepts in geostatistics is spatial autocorrelation that is used to analyze patterns and similarities of monuments in dependence on their proximity to each other. For example, the analysis may reveal that certain types of monuments are often located near watercourses, which may indicate the historical importance of these sites to the population. A suitable tool for modeling and simultaneous understanding the spatial structure of monuments is a variogram (Vaughn 2019, 4–5). Let us suppose we are investigating the distribution of historical settlements in a certain geographical area. The variogram could help us identify how quickly the similarity amongst settlements decreases with increasing distance, providing relevant information for their protection and conservation. Another advanced geostatistical modeling technique, known as kriging, can be used to interpolate unknown values in space based on known data about the position of monuments (Vaughn 2019, 14–16; Hoef 2001, 3). For example, if we have incomplete data on the distribution of archaeological sites, kriging can predict the likely location of another sites based on patterns obtained from existing data (Rzasa/Ciski/Ogryzek 2019).

Geostatistics can be applied to broader monitoring and planning of cultural landscapes, where it can help in creating effective strategies for the monuments management. For example, the analysis can identify regions with high density of historic monuments that may require increased protection from urbanization pressure or environmental threats. Thus, the significance of geostatistical analysis for cultural landscape monuments lies in its ability of comprehensive and quantitative understanding of geospatial patterns and processes. This information is crucial for the cultural heritage protection and sustainable management, thus allowing the landscape historical and cultural value to be preserved for future generations.

- **Spatiotemporal analysis** allows for effective study of dynamics and patterns of change of monuments in a historical landscape. This approach integrates temporal data into spatial analyses, which provides a more comprehensive picture of the evolution and interactions of cultural and historical objects across both space and time (Rząsa/Ogryzek/Ciski 2019, 1–2). In practice, spatiotemporal analysis can help identify degradation trends in historic structures, which may be influenced by different factors, such as climate change, urbanization pressure or tourist activities. For example, monitoring a land use change around medieval castles can reveal increasing risks from the infrastructure development or changes in the landscape architecture that may threaten their preservation. Another application of this analysis is monitoring of changes in the availability and condition of driveways to monuments (Rząsa/Ogryzek/Ciski 2019, 3). The spatiotemporal data can show historical changes in the accessibility, which can be important for conservation plans and sustainable management of these sites (Flamary et al. 2015, 5). An example can be the analysis of historical and current photographs and maps, which can reveal the gradual erosion of roads accessing important archaeological sites. The advantage of spatiotemporal analysis is its ability to include large amounts of data from different time periods, allowing researchers and conservationists to better learn the long-term patterns and predict possible future changes. This information is crucial in creating the management strategies that consider historical, cultural and environmental factors affecting the monuments.

- **Predictive modeling** makes it possible to analyze and predict future scenarios regarding the conservation and management of monuments in the historical landscape. The use of mathematical and statistical methods in predictive modeling allows formulating reasoned hypotheses about potential changes that may affect historical sites, thus contributing to their better protection and sustainable management. The models are based on a large number of examples and are used to predict results for new data (Lamba/Madhusudhan 2022, 213–214). There are various machine learning methods and algorithms that are applicable in GIS, such as support vector machines (SVM), decision trees, and neural networks. These algorithms make it possible to solve complex tasks in predicting and classifying the spatial data (Lamba/Madhusudhan 2022, 216–217). One example of how predictive modeling can help in the monuments protection is in analyzing the impact of climate change on medieval castles. The modeling can identify factors, such as changes in temperature and precipitation that could lead to accelerated erosion processes or other structural damage in the future (Johnston et al. 2003, 4). This information can then be used to plan the conservation measures to minimize these risks.

  Another area of the predictive modeling use is monitoring of urbanization pressure on archaeological sites. The method makes it possible to predict how the development of urban or industrial areas could affect monuments, which allows intervening early by means of legislation or planning to protect these sites. Predictive modeling can also provide valuable inputs for managing the number of visitors to historic sites. Analysing the past attendance data, together with anticipated tourism trends, can predict future peak periods, enabling better planning of tours, maintenance and conservation measures to minimize possible damage. However, predictive modeling can also bring some challenges, such as the need for accurate and extensive data components and complex analyses, which can correctly interpret dynamics and interactions within historical and cultural landscapes. It is also necessary to consider the uncertainties associated with each model, which may be resulting from limited or inaccurate input data (Rzasa/Ciski/Ogryzek 2019, 64–65). Each of these methodologies enables uncovering of patterns, trends, and causations in geospatial data. Some of the geospatial analyzes mentioned above can be demonstrated on selected simple model situations.

Model situation 1.

A simple spatial analysis of monuments layout according to their registration in the list of immovable cultural monuments in the area of Ipeľská kotlina hills[1]. The aim of this analysis was to define spatial areas with significant number of monuments registered in the cultural monuments list. The model analysis is the QGIS program output.

Step 1: Data preparation

Primarily, it is necessary to obtain geospatial data containing the locations of monuments in the selected region. The monuments were precisely located and their typology was made, based on the available list of immovable monuments (NNKP Register), which was supplemented by our own field verification of the monuments position and condition. Individual object points were created manually in a single vector layer.

Step 2: Analysis of the monuments density

Based on the Density Analysis program add-on, a raster map (Fig. 12.) was created in the form of a colour scale that analyzes the results and identifies the areas with the highest concentration of monuments. As can be seen on the attached map, most of the law-protected immovable monuments are concentrated in the eastern part of the area under investigation, i.e. in the area of Pôtorská pahorkatina hills and, to a lesser extent, in the area of

1 The term 'Ipeľská kotlina' refers to a hilly region in southern Slovakia. The original Slovak name is used in this book to preserve the geographic and cultural authenticity of the region, as well as to provide readers with its proper local designation commonly recognized in Slovak academic and geographic literature.

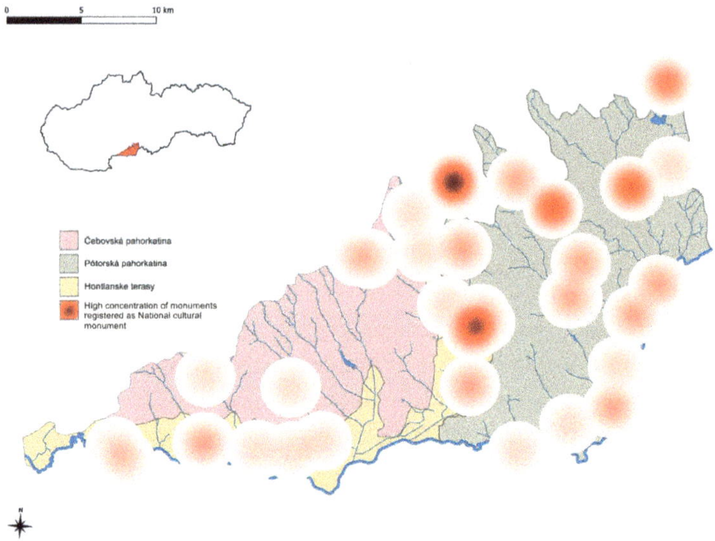

FIGURE 12

Hontianske terasy (Hont terraces). These areas may represent the regions of high historical and cultural significance that can require deeper protection of monuments or can be attractive for tourism, while at the same time reflecting the current state of the monument protection in the given region.

The given model situation can be supplemented by the same analysis, which will include objects that are currently not registered in the list of immovable cultural monuments, but have an assumed potential to represent the cultural-historical and landscape-historical development of the selected region. The mutual comparison will allow identifying of the territories where possible new research can be carried out, leading to the registration of new monuments, which would contribute to protection of a wider spectrum of cultural monuments in the given territory.

Model situation 2.

Additional analysis of selected im movable landscape-historical objects with the potential for deeper research and eventual registration in the list of

National cultural monuments on the same territory. The analysis was aimed in obtaining an overview of areas with more frequent architectural constructions, which will be compared to the results obtained in model situation 1. In addition to comprehensive analyses of all types of buildings, segmented analyses of individual types of objects were carried out, too. The model analysis was realized in the QGIS program.

Step 1: Data preparation

Geospatial points were created in the QGIS program according to data from the literature (Güntherová 1967, 1968 and 1969), map clients and field surveys. The points representing individual human-geographical objects were divided according to type-functional properties into the following segments: sacral objects, secular buildings, historical greenery, natural monuments, technical monuments, memorials, gravestones, cemeteries and archaeological sites.

Step 2: Analysis of the monuments density

On the basis of the same Density Analysis software add-on, a raster map was created which, in the form of a colour scale, analyzes the results and identifies the areas with the highest concentration of immovable landscape-historical elements (Fig. 13). For this process, it was necessary to merge all types of identified objects into one multipoint layer. After this step, the Density Analysis add-on was used. The resulting model image represents, according to the intensity of the colour scale, areas with a high, low or no concentration of immovable human-geographical objects. In this case, more remarkable occurrence of the objects under investigation is again within the area of Pôtorská pahorkatina hills and Hontianske terasy (Hont terraces).

The result of both model situations are colour raster maps with circles, the colour intensity of which indicates the occurrence density of the investigated elements. Comparing the both model situations shows that the number of potentially valuable landscape-historical and anthropogenic objects in the studied region is much higher than of those registered in the list of national cultural monuments. The monuments are concentrated especially in the area of Hontianske terasy and in the southern part of Pôtorská pahorkatina hills. Objects in the Čebovská pahorkatina hills are somewhat more frequently present as well, which in both cases is not sufficiently reflected in the number

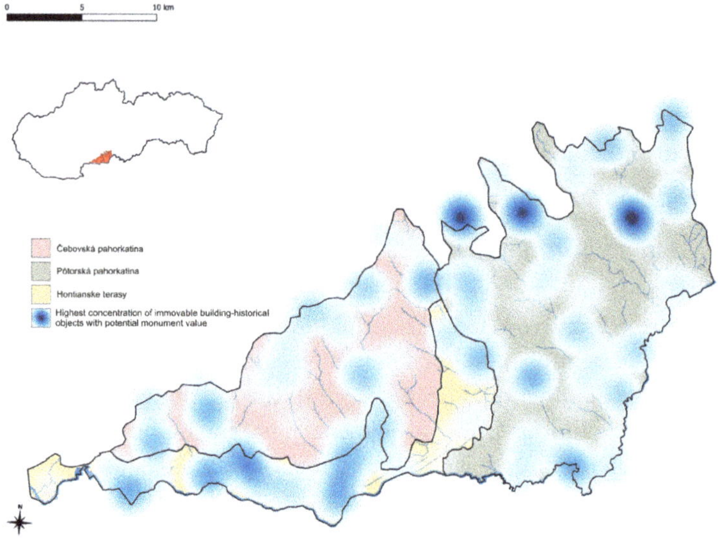

FIGURE 13

of law-protected and potentially law-protected monuments. There are several possibilities for clarifying our findings more precisely. Primarily, this situation is mainly caused by constant transformation of the landscape cover and intensification of its individual structures, which naturally changes their composition and, at the same time, historical anthropogenic objects as well. The landscape changes, caused by various situations, manifested themselves differently in the area in question, and this is probably the reason why the monument concentrations are higher in some places and completely absent in others. Further detailed model analyzes in the GIS space would certainly bring more accurate results.

7.1.2 3D modeling in GIS

This method represents a fundamental shift in the visualization and analysis of geospatial data. The use of three-dimensional imaging in GIS opens up new possibilities for planning, analyzing, simulating and presenting the geographic data in contexts where two-dimensional methods cannot adequately depict complex relations and environmental dynamics. 3D GIS can integrate digital terrain models (DMT) with vector and raster layers, thus enabling a three-dimensional representation of the earth's surface and the objects above. Terrain models are often obtained by techniques, such as LiDAR (Light Detection and Ranging) or photogrammetry, which provide high resolution and accuracy in the representation of terrain formations (Schulze-Horsel 2007, 1). To complement the terrain model, 3D models of buildings, infrastructure, vegetation and other objects are used. These models can be created manually by using 3D modeling tools or automatically generated from data obtained by drones or the other remote sensing techniques mentioned above. Integrating these data into unified 3D environment allows users to browse, analyze and interact with the model in a more intuitive and realistic format.

The basic principles and technologies of 3D modeling in GIS are focused on three main areas: creation of 3D models, processing and analysis of 3D data, and visualization of three-dimensional scenes. These aspects are the basis for understanding and effective use of three-dimensional geospatial information systems.

Creating 3D models:

- **Data acquisition:** The first step in the 3D modeling process is to acquire accurate geospatial data. The data can be obtained by various methods, especially LiDAR, photogrammetry, satellite or orthographic imaging that are mentioned in the third chapter.
- **Terrain and object modeling:** Digital terrain models (DTM) and digital surface models (DSM) are the basic building blocks for 3D models in GIS. DTM provides a detailed foundation of the terrain, while DSM includes above-ground objects, such as buildings and vegetation. The modeling process will be commented in more detail in the text below.

2. **Processing and analysis of 3D data**

- **Data integration:** 3D GIS systems must efficiently integrate different types of data from different sources and in different formats. This requires high-performance database systems and the right data processing tools.

- **Data analysis:** GIS software provides tools for 3D data analysis, which may include volume computation, visibility analysis, sound or light spreading modeling, and other three-dimensional analyses.

3. **Visualization of three-dimensional scenes**

- **Rendering and navigation:** Visualization of 3D scenes in GIS involves complex rendering techniques that enable realistic display of geospatial data. Users can navigate 3D scenes interactively, which improves the data understanding.
- **Interactivity:** Modern 3D GIS systems allow users to interact with the data through tools to rotate, zoom, and move. This interface can be improved with virtual or augmented reality, thus creating an even more immersive experience.
- **Performance and optimization:** Considering the huge amount of data involved in 3D GIS, to optimize the system performance is crucial. Techniques such as level of detail (LOD), where models decrease in complexity with increasing distance from the observer, are important for effective processing and visualization.

3D modeling in GIS brings numerous significant advantages in the geospatial data analysis and presentation. As technology advances, the capabilities of 3D GIS are expanding, thus allowing for a better comprehension and management of the physical world. These systems are constantly evolving, integrating advanced technologies and approaches to provide more exact and user-friendly tools for 3D geospatial analysis or custom reconstructive 3D modeling.

The knowledge given above is demonstrated in further model situations.

Model situation 3.

If necessary, geospatial data can be visualized in three dimensions in selected GIS-supported programs. This quality is also present in QGIS program, which can model the terrain in three dimensions from the DMR raster map through its own model. For this purpose, a situation was chosen where it was necessary to visualize the terrain remains of a medieval fortified settlement (Hunka/Nevizánsky/Ožďáni 2001, 74–76; Plaček/Bóna 2007, 201) in the cadastre of Malé Zlievce village (distr. of Veľký Krtíš). The spatial visualization goal was to verify the terrain configuration of a fortification remains and the driveway course.

Step 1: Data preparation

DMR 5.0 of Malé Zlievce cadastre was downloaded from the portal of the Office of Geodesy, Cartography and Cadastre of the Slovak Republic (hereinafter referred to as OGCC). In QGIS, the S-JTSK coordinate system (EPSG:8353) was set and the DMR raster layer with the same coordinates was imported. The inserted grid representing the entire cadastral territory was cropped as needed, as it was not necessary to work with the entire area to visualize the small fortification.

Step 2: 2D terrain visualization

The raster layer had to be visualized properly in graded colour shades that indicated altitude values in the terrain. Five colour classes were chosen to depict individual altitude ranges. In this way, several artificial structures, such as mounds or terrain depressions, were showed off.

Step 3: 3D visualization

The terrain prepared in this way could be visualized using the 3D map displaying program function. In the new QGIS window, after setting the parameters of input altitude data, resolution of the visualized raster and shading effect, a 3D model of the selected area was displayed, on which remains of the medieval settlement fortification were clearly outlined (Fig. 14). Subsequently, it was possible to display individual parts of the fortification from different views and perspectives, which contributed to a deeper identification of anthropogenic structures and the overall character of the fortification.

The above presented model situation in the QGIS program does not represent a special research analysis, but rather a visual aid in interpreting or better identifying terrain structures. In QGIS, it is not possible to directly model 3D terrain, or to complete or modify it according to the requirements. The program has not been developed for this purpose, however, if modeling in GIS is needed, this program can be combined with other 3D

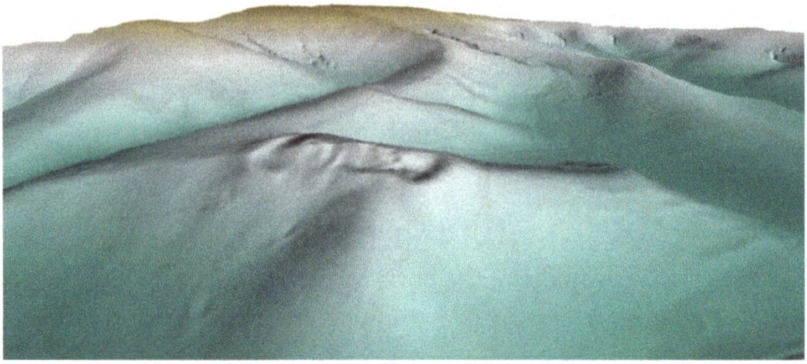

FIGURE 14

modeling programs, such as Blender. The possibility of interconnecting the above-mentioned programs is demonstrated by following model situation.

Model situation 4.

For this model situation, the modified DMR of the fortified settlement near the village of Malé Zlievce from the previous example will be used. In this case, the goal is to obtain a polygonal model of the selected territory in a format that can be imported into Blender program, in which the desired parts can be modeled.

Step 1: Export of 3D format (.stl) from QGIS program

DEMto3D program add-on was applied to prepared cropped DMR raster, which must be installed from the plug-in menu. After opening the module, DEM 3D printing window opens. By setting the required parameters, such as scale, model resolution and model size, it can be exported to STL format.

Step 2: Import into Blender program

The exported file with the extension .stl is imported into the program (Fig. 15). The terrain model can be modified in Blender according to the requirements in many ways. It should be noted that Blender provides

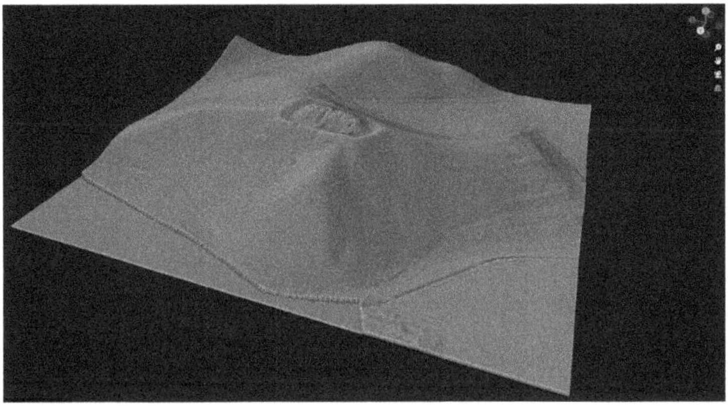

FIGURE 15

much better and more visually usable display modes than QGIS. In the case it is possible to reconstruct some parts, our own additional or evocative models can be inserted on the imported terrain. At the same time, if possible, past and present landscape cover or extinct but verifiable and substantiated landscape components can be reconstructed. In addition, several development phases of the landscape can be reconstructed on one model (Fig. 16). Finally, real sunlight and atmospheric conditions can be imitated in Blender.

This way of processing 3D data represents a very promising and simultaneously useful way of interpreting the scientific data. However, the exactness and quality of 3D model display in Blender is also depending on the quality of exported DMR raster map. When modeling monuments and landscapes, much smaller parts of the landscape or above-ground elements, such as buildings or vegetation, are often needed to be available for reconstruction purposes. The DMR raster represents already modified LiDAR data, where there is only one classification class left – the surface. In this case, the cloud of points obtained by the LIDAR method have to be used, which could, if necessary, be supplemented by the UAV aerial photography. The following model situation describes processing of a point cloud into a polygonal model.

FIGURE 16

Model situation 5.

The Cloud of points in the S-JTSK coordinate system, which occupies a smaller area of Dolná Strehová village, was downloaded from the OGCC portal. The subject of processing will be the former historical park with a mansion dated to the middle of the 19th century. This example will be used to demonstrate how to adjust a point cloud according to your own requirements and create your own 3D surface and land cover model or DMR for further processing or modeling. On the prepared model, it is then possible to reconstruct the original layout of the buildings, the mansion material form and the vegetation cover in the original park according to the period plans and pictures.

Step 1: Data preparation

The downloaded point cloud in the form of multiple files with the .las extension was imported into the CloudCompare program. Embedded files have been combined into one compact cloud for better handling and settings. During the import, the geospatial coordinates were converted to the local coordinates of CloudCompare, which, however, had no significant impact on the cloud scale.

Step 2: Visualization of the point cloud

In the CloudCompare program, the point cloud was visualized based on scalar fields according to the classification classes of LiDAR data (Table 1), in which each class can be displayed in a different colour. Better accentuation of the classes was achieved by choosing a higher contrast in the Colour Scale – High contrast setting (Fig. 17.). The cloud has clearly visible objects and now it is possible to check whether the objects are well classified in the classification classes.

Step 3: Separating the point cloud classes

To demonstrate this situation, the cloud was split and class 2 (surface) was selected, using the Filter points by value tool in the CloudCompare program. No other cloud parts were needed in this model situation.

Step 4: Creation of a polygonal terrain model

The point cloud that represents the terrain surface had to be defined by normals so that the program could correctly verify the orientation of the polygons in the future model.

After that, a detailed polygonal network based on triangular polygons will be created using the Poisson Surface Reconstruction program module (Fig. 18).

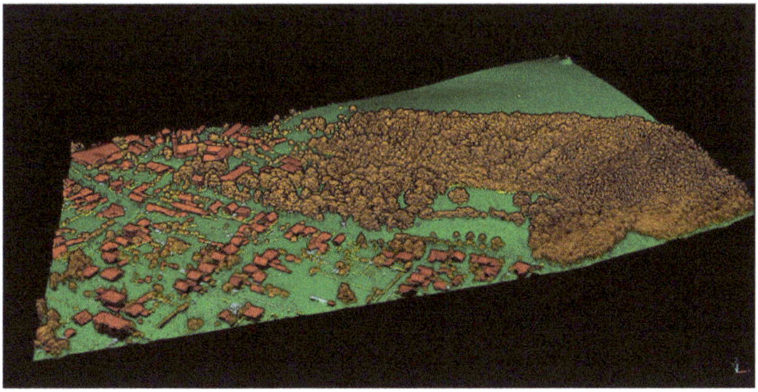

FIGURE 17

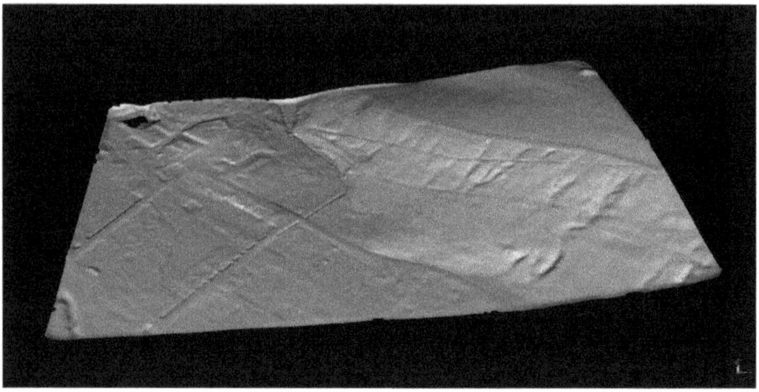

FIGURE 18

Step 5: Export of a polygonal model

Since our own 3D entities, such as the original buildings, urban structures or vegetation cover, cannot be modeled in CloudCompare, this spatial polygon is needed to be exported in a selected format. Here, the .obj format was chosen, which is multi-platform and therefore can be used in many modeling programs.

As can be seen from the selected model situations, currently there are numerous software tools and options for modifying or modeling spatial landscape models. The variability of multi-platform 3D formats makes it easy to combine multiple programs with GIS and modeling support. However, the development of some modeling programs is so exponential that it is possible to work with real geospatial coordinates within one program and at the same time to model, simulate and interpret the mutual contexts of monuments and historical landscapes. Such possibilities are offered for research and presentation by the Blender program, which has its own GIS module (Fig. 19).

Model situation 6.

This model situation works with the possibilities of effective, quick and relatively exact displaying of a rather large geographical space in a modeling program with some spatial entities, such as buildings, watercourses or roads. For this purpose, the GIS module will be used, which can be installed in the

FIGURE 19

Blender program. The goal is to map the territory of Veľký Krtíš town, to visualize the terrain geomorphology, spatial distribution of buildings and urban units and subsequently analyze and simulate individual development stages of the town from the end of 19th century in 3D space.

Step 1: Importing data into the Blender program

Through the GIS module offered in the program menu after the installation, a satellite map of the entire territory needed for analysis and simulation is displayed. Pressing the E key, only the selected area remains on the program desktop.

Step 2: Creation of 3D landscape relief

Using the GIS module, spatial parameters were chosen for the selected town area of Veľký Krtíš based on the OpenTopography SRTM 30m service. A 3D terrain has been created that can be further refined using the Subdivision Surface modifier or increased up to the Z-axis as needed with another Displace modifier from Blender menu.

Step 3: Importing of selected geospatial objects

In the GIS module, the required anthropogenic elements were selected through the Web geodata menu. In this case, they were the buildings forming the current development (buildings), waterways and highways. After the import, all required spatial entities appear as polygonal models or as line objects. It should be noted that in the case, building height differences are not included, as the automatically imported 3D building polygons are recalculated from the original 2D coordinates (Fig. 20). For more detailed modeling, each building can be modified individually as needed. This model situation does not consider it, as the goal is to visualize only the town territorial development and not the mass of the urban development.

Step 4: Visualizing the urban development

In this step, individual town districts were chosen and colour-coded according to the phases of built-up area development. The determination was based on 20th century maps and published sources (Hronček/Gregorová/Herčko 2021, 448). This modeling allowed, according to the mapping, differentiate the development of the street network in the 20th century and identify the original buildings of the town, or rather the village, in the 19th century. For the needs of final outputs, several visuals of the terrain foundations were created (Fig. 21.). In addition to the satellite base, it was an altitude model based on the colour-coded altitudes (Terrain analysis).

FIGURE 20

FIGURE 21

7.1.3 Map-based virtual 2D reconstruction

This method represents a relevant part in the field of cultural heritage digitization, where historical or archaeological maps are transformed into digitized formats to provide a detailed view of the cultural monuments history and development in the historical landscape. The process of virtual 2D reconstruction involves several key steps, with digitization of historical maps in the beginning. The sources of these maps can be various, including libraries, archives or other research institutions. After digitization, the maps are georeferenced, which means that the mostly digital raster maps are aligned with current coordinates, which allows them to be accurately compared to modern maps and geospatial data.

In the next step, digital models based on these georeferenced maps are created that can illustrate the terrain historical features, location of monuments and other relevant positional components. For example, the reconstruction can reveal location of a medieval castle and its surroundings, thus providing valuable information about its defense systems and logistic routes, which are often hidden in the current cultural landscape.

These reconstructions are particularly valuable in interpretation of archaeological sites, where combination of historic maps and contemporary archaeological research can be used to create rather exact reconstructions of past settlements and their interactions with the natural environment.

Model situation 7.

This model situation illustrates the method of effective, fast and relatively accurate reconstruction of a large geographic area of the wine-growing landscape in the territory of Čebovská pahorkatina hills in a historical context. The use of geographic information systems (GIS) in combination with historical maps and modern technologies, such as LiDAR, provides a detailed overview of historical changes in the vineyard landscape and enables exact display and analysis of the defined areas.

Step 1: Data preparation

Relevant historical maps from the 18th and 19th centuries were selected that show the wine-growing areas and their configuration. Maps of the first and second military mapping provide a basic visualization.

Step 2: Geographical comparison

Using the QGIS program, the old maps have been compared to modern topographic and orthophotographic maps, which allowed identifying of preserved structures and changes in the landscape configuration.

Step 3: LiDAR identification of old vineyards

LiDAR data (DMR 5.0) were used for detailed terrain scanning, which also revealed hidden anthropogenic structures and reliefs related to the historical vineyards (Fig. 22).

Step 4: Creating own map layers

Based on the obtained geodata, vector layers representing historical vineyard areas during different time periods were created. It enabled to compare and calculate the changes in the vineyards area (Fig. 23).

Step 5: Field verification of anthropogenic structures

Identified areas and objects were verified in the field by using non-destructive methods, such as photodocumentation and photogrammetry, which can support the verification of digital data and correctness of the cartographic analysis.

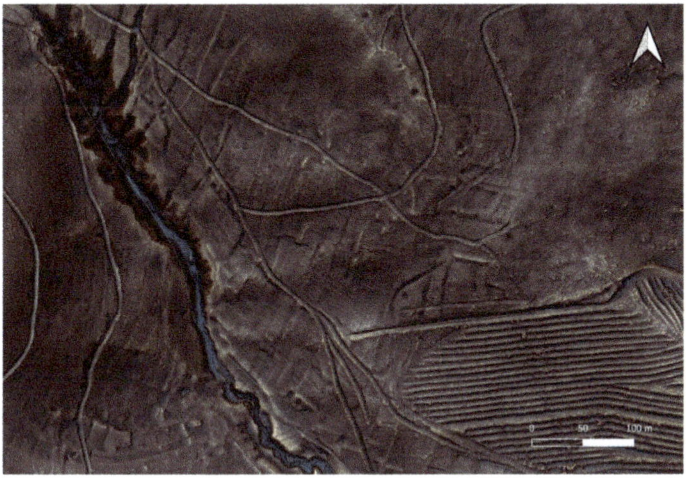

FIGURE 22

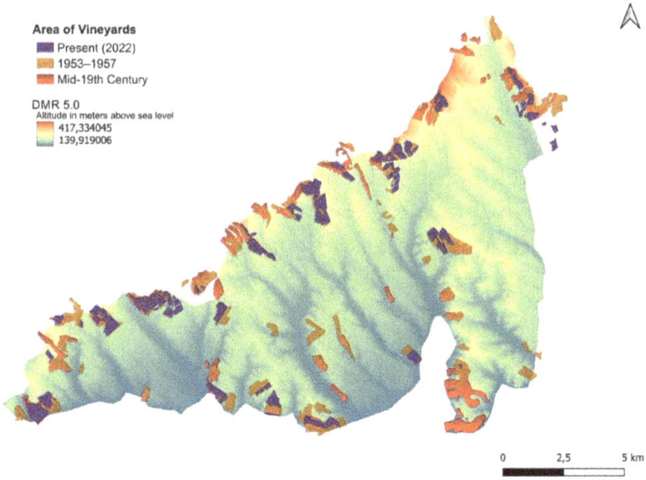

FIGURE 23

7.1.4 Visibility analysis and simulation

Visibility analysis, also known as sightline analysis or viewshed analysis, is the process of determining which areas are visible from a certain viewpoint point (Zhang et al. 2014, 4–5). In the context of monument protection, this analysis can help set from which points of a town or natural environment the monuments are clearly visible. For example, a visibility analysis can show that a certain historical structure is well visible from popular tourist routes, which is important for its tourist attractiveness and educational potential.

To carry out this analysis in QGIS, urban planners can use digital terrain models (DMT) along with data of buildings and vegetation height to obtain an accurate visibility model for the site. By setting the elevation of the observer and the analyzed point, it is possible to simulate different scenarios of how changes in the built-up area will affect the visibility of monuments.

Simulation in GIS allows modeling and predicting the results of various planning and conservation scenarios. Regarding monuments protection, simulation can provide valuable information about how new construction activity or changes in landscape structure will affect their visibility, accessibility, or even physical integrity. An example would be the simulation of the impact of high buildings construction in the vicinity of medieval castles, which could result in a reduction of their dominance in the landscape panorama. These simulations are also invaluable in assessing the impact of climate change on monuments. Modeling the change in the water regime due to climate change can, for example, help identify potential threats to archaeological sites in lowlands that could be flooded due to rising water levels.

Visibility analysis and simulation are crucial for the sustainable protection and development of monuments in historic landscape. They provide significant information that can help plan locations of new objects, roads or other interventions into landscapes.

Model situation 8

Visibility of execution places and gallows standing on them was an important aspect of historical landscapes. The fact that these sites were and still are clearly visible is directly determined by their location in the field. Their visibility from driveways, urban areas and other hilly locations in the vicinity was essential. Therefore, the visibility analysis is crucial in identification or verification of these objects. Considering this situation, two gallows in Čebovce and Modrý Kameň (distr. of Veľký Krtíš) were selected, visibility of which was analyzed in the QGIS program. At present, these objects are quite difficult to locate in the field, as they were very simple structures. Therefore, visibility analysis can bring more fundamental information about more exact positioning based on the view from the surrounding landscape. In one case, the so-called reverse visibility analysis was used as well. This method is useful when the exact location of the gallows is not known and there are only positions and roads selected for the visibility recalculations. In the visibility intersections the place of executions can be identified more accurately.

Step 1: Data preparation

DMR 5.0 maps of the selected locations surroundings were imported into the QGIS program. Then the maps of the first and second military mapping from the end of the 18th to the half of the 19th century were georeferenced manually (according to the possibilities of historical maps).

Step 2: Creating vector layer

To calculate the visibility, it is first necessary to create a point from which it will be calculated. A point as a vector layer was created on the places where, according to historical mapping, the execution site is assumed. The location is still approximate, as the deviation of the historical maps from the current ones can vary even by several tens of meters.

Step 3: Creating the target and observation point visibility

In the QGIS add-on called Visibility analysis, the location parameters were set first, from which the visibility will be calculated. In particular, the target point height, the observation point height and the radius in which the calculation is to be carried out are entered. In the case of gallows, the height could range up to five meters above the surface. The observation height based on the average height of a person (approx. 1.65 m) was entered as well. The calculation radius should exceed the entire urban area and most of the extra-urban area of the settlement. Then the visibility is calculated using the Viewshed function (Fig. 24).

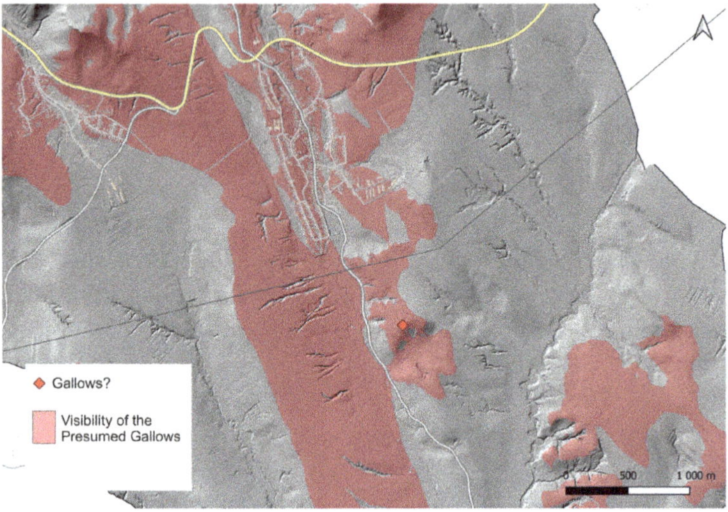

FIGURE 24

Step 4: Reverse visibility analysis

The method is identical with the classic one, but in this case, multiple point vector layers on selected points in the terrain were created from which the visibility will be calculated. This means that the points are placed at points from which the gallows should be visible (urban area, roads, nearby elevations). Each point is assigned height parameters and calculated view radius. The most suitable place for situating the gallows can be chosen from several intersections of surface visibility interpretations.

Step 5: Output export

After generating the raster images that interpret the visibility in the field, their colour and transparency can be adjusted to better highlight or distinguish the information required. It is necessary to point out that possible inaccuracy or relative results have to be verified in the field (Fig. 25).

7.1.5 Multitemporal layers

Multitemporal layers are crucial for understanding how historical and cultural sites have evolved under the influence of natural and human factors. They

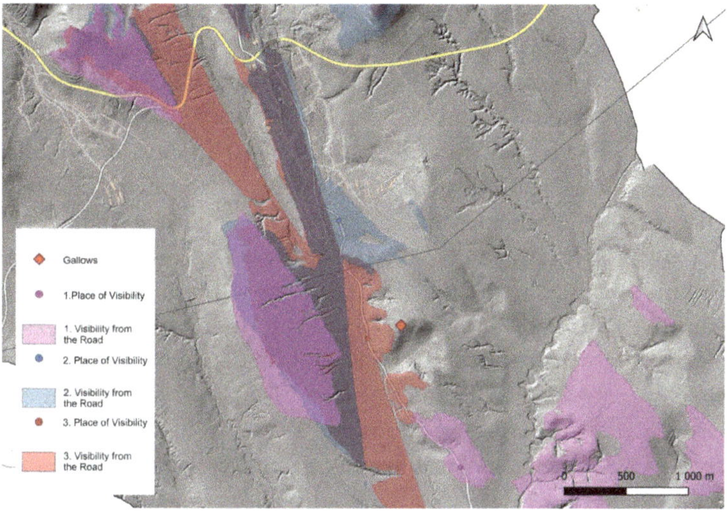

FIGURE 25

provide a unique implement for comprehending the historical processes that influenced the development of cultural and historical monuments (Dadras et al. 2015). For example, the analysis of satellite images acquired over different decades can reveal gradual urbanization around important archaeological sites, leading to changes in their surroundings that can threaten their integrity and condition. In practice, multitemporal layers are used to monitor changes in monuments condition and in their immediate vicinity. Using GIS technology, the way how driveways, vegetation or waterways around historical castles have changed can be monitored. This information is invaluable in planning the monument conservation works and for the administration responsible for the protection of these sites. Multitemporal analyzes in GIS are technically demanding because they require processing of large volumes of data and their integration into coherent layers. The data have to be exactly georeferenced to ensure reliable results (Michele et al. 2019). Different data formats, differences in data quality and availability for different time periods represent other challenges that need to be solved (Flamary et al. 2015, 4–5).

The use of GIS methods opens up new possibilities for visualization and simulation in fields such as monument care, cultural heritage protection, environmental monitoring and others. Visibility analysis, multitemporal layers and other mentioned methods are basic methods for a deeper understanding of geospatial data, which are necessary for creation of accurate and information-rich 3D models. However, when using these methods, it is important to keep in mind that a successful application requires not only the correct selection of methods, but also a thorough understanding of each technique limitations and precise processing of input data. As with any analytical method, exact results are directly dependent on the quality and accuracy of the data used. These methods allow users to identify and analyze patterns, trends and dynamics of the landscapes or objects under investigation, which is crucial for successful implementation in specialized 3D modeling tools such as Blender or CloudCompare. Integration of these tools enables transformation of traditional 2D data into dynamic 3D models that can offer better understanding and interaction with geospatial data.

In this chapter, only a few selected GIS methods have been presented, which represent a small selection from the vast spectrum of available tools and techniques in the field of geospatial analysis. Although there are many more methodologies and applications that can be used in different contexts and disciplines, this set of methods was chosen for its relevance and applicability in the context of 3D modeling in specialized programs. GIS modeling is increasingly being combined with advanced 3D graphics programs such as Blender and CloudCompare, which enables more accurate and detailed visualizations of geospatial data. The mentioned software tools provide extended options for analyzing and manipulating 3D models that exceed the traditional capacities of standard GIS programs.

For example, Blender can be effectively used to create complex visualizations of geographic data. Due to its ability to import various data formats, including those from GIS platforms, Blender can be used to create realistic 3D scenes of geographic sites. Blender models can contain detailed terrains, architectural structures and dynamic scenery, which is ideal for visual presentations and landscape simulations. On the other hand, CloudCompare is a specialized tool primarily focused on the processing and analysis of point clouds and 3D meshes, which data are often obtained from LiDAR or photogrammetric measurements. CloudCompare provides users with robust tools to register compare and visualize point clouds, enabling detailed analysis of physical characteristics of terrain or objects. These capabilities are crucial in monitoring environmental change.

7.2 3D modeling in specialized software based on polygonal models

3D modeling of monuments and historic landscapes of structures is a complex process involving different approaches and techniques. 3D modeling of historical monuments includes geometric, topological and semantic modeling, which combines aerial and architectural photogrammetry (Nour El Din/ Grussenmeyer/ Koehl 2000, 1). This approach enables accurate and detailed modeling of monuments, which is crucial for their protection and management. Modeling is not only about creating digital copies of real objects, but about capturing their history, structure and aesthetics in digital form. This process requires both technical skill and a deep understanding of the modeled object. Objects are structured into various elements of smaller topological dimensions, which allow for continuous updating of content and precise modeling of monuments (Nour El Din/ Grussenmeyer/ Koehl 2000, 8). One of the basic ways of 3D modeling is creation or modification of polygonal models. Polygons, the basic building blocks of 3D graphics, are made up of vertices, edges and faces, that work together to shape a geometric model. Each polygon is typically a triangle or a quadrilateral, and their correct arrangement and number determine the model particularity and quality (Fig. 26.). In the context of historical monuments, polygonal modeling is particularly important because it allows for a high degree of detail while maintaining a relatively low demand on computing resources. The data obtained by selected digitization procedures are then processed into three-dimensional points (point clouds) that define the shape of the object. The point clouds are subsequently converted into polygonal models. This process can include various techniques such as retopologization that can simplify and optimize the number of polygons without losing details, which is essential for archival and presentation purposes of monuments (Sánchez 2024, 7).

In the context of monument care, 3D digital modeling is seen as a revolutionary technology that is transforming the way we approach the documentation, conservation, restoration and presentation of cultural heritage. 3D modeling makes it possible to capture extremely detailed images of monuments,

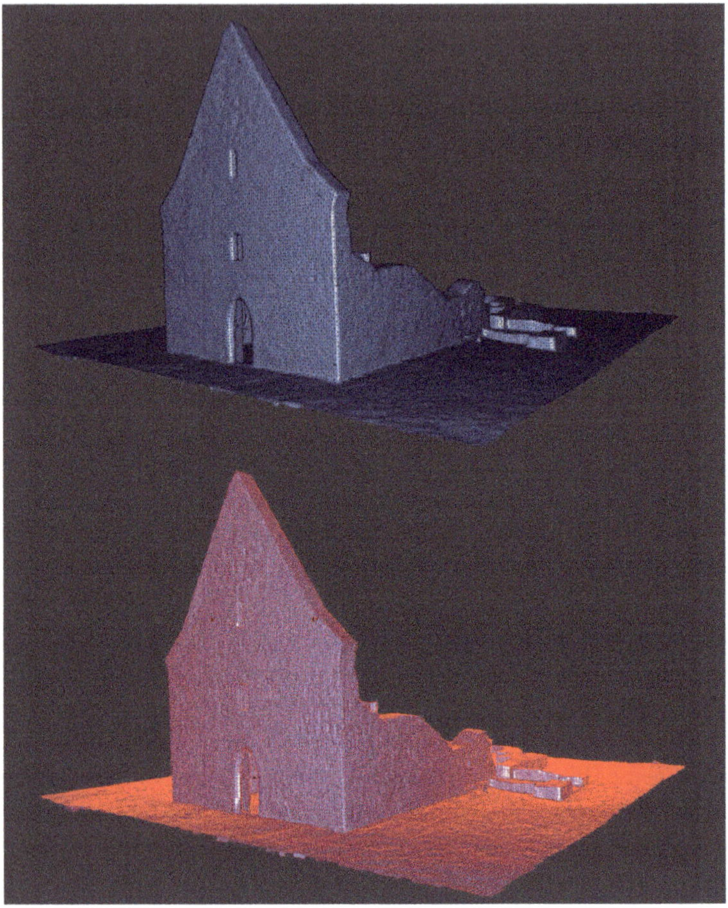

FIGURE 26

which is beneficial for both visual presentations and analytical purposes. Models created using laser scanning or photogrammetry can capture subtle details of architecture or art that might be missed with traditional methods. Modeling in virtual space brings several important possibilities when, for example, digitized objects can be complemented and enriched with hand-created virtual segments based on our own erudite research of the given issue. Many, especially immovable monuments in situ or historical landscape structures, have remained only in fragments, and their physical "reconstruction" is often very difficult or even impossible. This serious deficiency can significantly eliminate the work of a virtual space, in which de facto there are no spatial barriers or restrictions. Of course, it should always be kept in mind that these entities are unreal and therefore do not exist in their physical essence.

Currently, there are plentiful programs that are offering a large number of possible tools and functions for creating, editing and visualizing digital models. Their role is crucial in several aspects:

1. **Digitization of real objects:** 3D modeling software allows you to transform physical data obtained by scanning or photography into detailed digital replicas. These programs provide tools for processing raw data, such as point clouds, into complex 3D models with precise geometric shapes.
2. **Model optimization:** In monuments, it is often necessary to optimize models to be suitable for digital archives or interactive applications. Modeling programs such as Blender, 3ds Max, or Maya can offer extensive options for retopologizing, which reduces the number of polygons in the model without losing details. This will ensure that the models are visually attractive while efficiently using computational resources.
3. **Texturing and materials:** To achieve realistic appearance of historical objects, applying textures and materials correctly is important. Modeling programs provide advanced tools for texture mapping and simulation of materials, which allows accurate display of surface characteristics such as stone, wood, or metal.
4. **Visualization and rendering:** To present and analyse models, they have to be visualized adequately. 3D modeling software includes extensive rendering capabilities to create photorealistic images and model animations. These functions are substantial for presenting models to scientific communities or to the public.
5. **Interoperability and export:** Modeling programs play an important role in sharing and integrating 3D models into various digital platforms. They support export to standard formats that are compatible with other

tools and systems, thus allowing for easier collaboration and accessibility of models.

3D modeling of monuments and historical landscapes is a complex and multidisciplinary process that includes various methods and techniques digitally capture, restore and visualize historical objects. This process enables not only accurate documentation and conservation of monuments, but also their detailed study and presentation to the public in a new perspective. Development of applications that allow interaction with 3D models via PC or mobile devices is crucial for the cultural heritage propagation and for education of the general public (Sánchez 2024, 8).

Composing of 3D virtual scene

At the beginning of every virtual 3D project, there is a virtual space into which model entities are gradually inserted, modified or newly created. This space can be termed 3D scene. Virtual 3D scene is digital representation of a space or environment that can be designed to simulate real-world situations or completely fictional worlds. These scenes are cornerstones of many creative spheres, including video games, movies, virtual reality (VR), architecture, and training simulations (Fig. 27, 28).

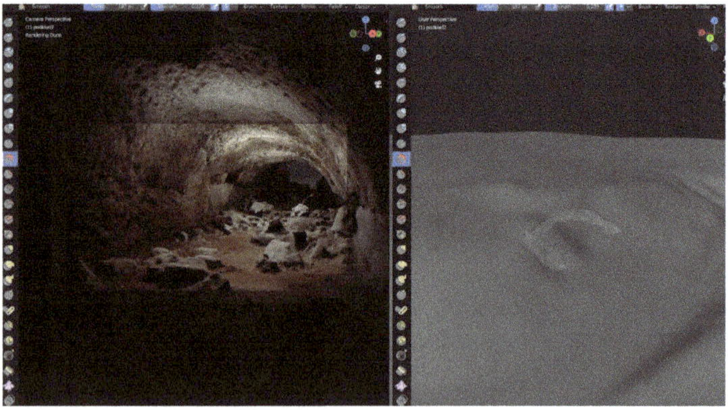

FIGURE 27

FIGURE 28

Virtual scene can consist of multiple virtual entities that are based on the 3D modeling concept. There are key modeling concepts, such as virtual anastylosis, evocative models, hybrid models, etc., that are integral part of this process and it is important to apply them to the procedures of a virtual scene creating (Forte 2008).

Virtual anastylosis

Virtual anastylosis is a process in which architectural structures or artefacts that were damaged or destroyed in the course of history are digitally reconstructed and preserved in the form of a torso. The term is derived from traditional anastylosis, a technique used in archaeology and conservation to physically reconstruct ruins by combining original parts with new materials to restore the original shape of the object. Virtual anastylosis differs in that the entire process takes place in a digital environment, which adds new possibilities and challenges.

The key advantage of virtual anastylosis is its ability to reconstruct historical objects without physical interference with existing artefacts or structures. This is especially useful in cases when original materials are too fragile or valuable and any manipulation would pose a significant risk. The process begins with detailed scan of existing remains or fragments using technologies such as laser scanning or photogrammetry, which can record their exact dimensions and shape. After obtaining this data, digital models are created and edited in specialized 3D modeling programs, where the fragments can be examined, modified and recomposed into their original or presumed shape. This process allows scientists and conservators to experiment with different configurations before finding the most likely reconstruction, which would be difficult or even impossible in real anastylosis.

Virtual anastylosis also provides significant benefits for education and presentation. Digitally reconstructed objects can be displayed in interactive exhibitions or online, allowing a wider audience to understand the history and cultural significance of the artefacts without travelling to remote or inaccessible sites. Also, it offers archaeologists and historians the opportunity to share their findings and hypotheses with the international scientific community, thus supporting the international cooperation and exchange of knowledge.

However, the reconstruction accuracy depends on the input data quality and scope as well as on subjective decisions in the fragments interpretation that can affect the resulting model. Discussions on methodology and standards are therefore an integral part of this field in order to ensure the reconstructions are as objective and accurate as possible.

Evocative models

In the context of 3D modeling of monuments and landscapes, they represent a specific approach that is focused on creating an emotional and interpretive representation of historical objects or scenes. These models are not focused first on exactness and technical perfection as it in virtual anastylosis, but rather on evoking the feeling, atmosphere and historical context of a given place or object. Evocative models are mainly used in educational presentations, where it is important to engage the viewer and provide him with a deeper experience of visiting the historical site or getting to know the object. In museums or virtual exhibitions, for example, these models can help visitors better comprehend the life and culture of the people who lived in a given place in the past.

Evocative models often integrate visual and audio effects that aid immersion and create the feeling that the viewer is part of the time being presented.

Creating an evocative model requires both technical skills in the field of 3D modeling and a deep understanding of the historical and cultural context of the modeled object. The scene creator has to work with historians, archaeologists, and other experts to ensure that the model reflects correct historical conditions. In addition, it is necessary to consider how various elements of the model, such as textures, lighting and interactivity, will contribute to the overall emotional impression.

Hybrid models

These virtual models combine multiple technologies and methodologies to create complex and multifunctional digital representations. These models connect traditional 3D modeling techniques with modern digital and simulation technologies such as augmented reality (AR), virtual reality (VR) or geographic information systems (GIS). The main goal of hybrid models is to increase the realism, interactivity and information value of 3D scenes. In practice, this can mean, for example, the integration of photorealistic 3D models of architectural monuments with historical data and multimedia content, which allow users to deepen their understanding and experience when interacting with the model.

Hybrid models are often used in areas where complex systems have to be displayed and analyzed in detail, such as urban planning, environmental studies or cultural heritage. By combining different types of data and visualization techniques, these models can offer more complex and flexible basis for analysis and decision-making than traditional one-dimensional models.

In addition to the technical integration of different software platforms and data resources, hybrid models require a multidisciplinary approach. Collaboration of experts from different fields such as architects, programmers, designers and historians is essential for creating accurate and functional models that can serve their purpose effectively.

Holistically reconstructed models

These are models that pursue providing a comprehensive and holistic picture of the object or scene being modeled. This approach includes physical aspects of the object and also its historical, cultural, and contextual attributes, thus striving for the most veracious and comprehensive digital representation possible. Holistically reconstructed models are often used in archaeology, history and cultural heritage, where it is important to represent the object physical appearance accurately and at the same time to understand its functions, meaning and place in a historical context. These models can include simulations of original materials, colour schemes, even sound and lighting conditions that can bring the viewer closer to the period milieu.

Creating a holistically reconstructed model begins with in-depth research and analysis of available historical data, including photographs, maps, written documents and physical remains. This research helps determine how an object or scene might look and function in its original conditions.

Subsequently, 3D modelers and professionals use specialized software to create digital models that integrate the knowledge gained. During this process, the techniques such as photogrammetry or laser scanning may be needed to capture exact physical details, which are then supplemented with textures and materials chosen to match the original as closely as possible. Creating holistically reconstructed models often requires collaboration of historians, archaeologists, architects and 3D graphic designers. Each of them contributes their specific knowledge and experience, ensuring that the model is both visually accurate and historically and culturally authentic.

Behaviour and organisms

In 3D scenes, behaviour and organisms refer to simulation of living beings and their interactions in digitally created environment. This includes not only the animation of movements and visual manifestations, but also the complex modeling of their biological, ecological and social behaviour. Organisms modeled in 3D scenes behave according to predefined rules or algorithms that simulate their natural behaviour in reality. This behaviour is usually controlled by scripts or artificial intelligence. The algorithms used can involve simple periodically repeating systems, complex conditions and transitions, or even machine learning to make characters adapt and respond to changes in their environment.

Furthermore, these models have to be equipped with a "brain" that enables autonomous decision-making without constant intervention from the user or programmer. This demands deep understanding of biology, ethology and other relevant sciences to make the simulations convincing and authentic.

For example, when reconstructing historical monuments, adding modeled organisms such as humans or animals can make the scene much more vivid and engaging. When modeling the landscape, animal models can be added to increase the authenticity and visual appeal of the scene. Patterns of bird migration, herds of animals, or even seasonal changes in vegetation can add a deeper and more dynamic feel to an environment that might otherwise appear static and lifeless.

Behaviour and organisms in 3D scene represent a complex combination of graphic design, programming and biological expertise. Integrating them into virtual environments can extend the user's experience, provide educational tools and expand creative possibilities for the content creators.

Virtual landscape segments and landscape cover

Virtual landscape segments are digital representations of real or fictitious territories that are modeled and simulated in 3D environment. These segments represent sections of the landscape with all their characteristic components, such as terrain, vegetation, waterways, roads, and other objects that define it. Virtual landscape segments are aimed to provide a realistic and interactive visualization of geographic and ecological systems.

In creating virtual landscape segments, advanced 3D modeling software and geographic information systems (GIS) are used. These tools allow integrating of various types of data, including satellite imagery, topographic maps and environmental databases to create high-leveled and exact details. The process involves digital terrain modeling, applying textures that represent different types of surfaces, and adding 3D objects such as trees, buildings or roads. One of the main advantages of virtual landscape segments is their interactivity. Users can move around the virtual landscape, observe it from different angles and interact with individual elements. This complex and highly interdisciplinary process gives possibility to reconstruct and visualize historical landscapes with high degree of detail and precision.

The above-mentioned virtual 3D scene components, due to their unique properties and applications, provide immensely rich opportunities for numerous areas of research and development. These technologies offer facilities for accurate simulating and visualizing of different scenarios, which have a direct impact on effective decision-making and planning.

Despite the many advantages, there are also challenges associated with the creation of virtual landscape segments, such as ensuring the exactness and updating of data. Future development in artificial intelligence and machine learning can promise further improvements that could automate and streamline the entire modeling process. These segments are thus a key tool that transforms the way we learn and interact with our physical world, opening up new possibilities for analyzing, educating and planning in the digital age.

When modeling historical objects, to consider textures and materials that can reproduce visual properties of objects is essential as well. These are often extracted from photographs and digitally edited to closely match the actual material. 3D modeling software offers a wide range of tools for texture mapping and simulating of lighting effects, which contribute to the model realistic display.

Visualization of materials and textures

In 3D modeling process and digital graphics, visualization of materials and textures is a fundamental step, which significantly affects the visual authenticity and aesthetic quality of digital objects and scenes. This technique is crucial in creating realistic representations of various surfaces and materials, such as metals, textiles, stone, wood and many others. Effective material visualization is based on understanding the physical and optical properties of materials that have to be digitally simulated. Each material has its specific qualities, such as reflectivity, surface roughness, transparency and texture, which need to be represented accurately. The modeling process begins with creating the object 3D model, to which textures and materials are subsequently applied in special 3D modeling software, such as Autodesk Maya, Blender, or 3ds Max.

Textures are digital images that are mapped onto the surface of 3D models to imitate the visual details of real ones. Texture visualization includes several techniques:

- Bump mapping increases the surface visual complexity by simulating bumps
- Displacement mapping changes the model surface geometry according to its texture, thus creating real depth and relief

- Specular mapping determines how light rays are reflected from the surface, affecting gloss and reflections.
- Normal mapping is similar to bump mapping, but provides a better simulation of small details without the need of geometry change.

Once materials and textures are applied, a rendering process follows, which can be done in real time or using ray tracing techniques. Real-time rendering is often used in video games and interactive applications, as it provides a good compromise between display quality and speed. Ray tracing, although more intensive in computation, produces remarkably realistic images with exact modeling of lighting effects, such as shading, reflections, and global illumination.

Precise visualization of materials and textures is critical not only in the entertainment industry, but also in architecture, interior design, simulations and education. In architecture, accurate visualization of materials can help customers better understand the designed space before it is actually built. In the field of education, realistic visualization can help lay public better understand anatomical models or historical reconstructions.

Regarding the historical monuments reconstructions, to achieve a high level of authenticity and visual realism is crucial. Properly applied textures can help restore the original appearance of materials, such as stone, wood, metal, or plaster that may have been changed or deteriorated in the course of time. Textures can cover everything from erosion and weathering to historical material processing techniques, thus increasing the reconstruction veracity. Visualizing the materials can provide valuable information about construction techniques and materials used in the past. For archaeologists and conservationists, material simulation can be a tool for understanding how historical objects were built, what techniques were used and what materials were available at any given time. Digital modeling and texture visualization can be used to document and archive the monuments condition before their restoration or before natural disasters. These models serve as a digital record that can be used for study or as a reference point for future restoration works.

Visualizing textures and materials can help professionals in better planning of restoration works. Simulations can predict how different materials and preservative techniques would possibly affect the monument overall appearance and durability, which can lead to more exact and sensitive approach to its conservation (Fig. 29, 30).

Effective visualization of textures and materials requires access to top software and hardware. The programs used for 3D modeling and rendering have to be powerful enough to handle the complex textures and materials that are specific for historical monuments. Precise visualization of textures and materials starts with high-quality data. This means that it is necessary to have access to accurate measurements, photographs, perhaps even historical

FIGURE 29

FIGURE 30

documents that describe the original materials and techniques used to build the monuments. Photogrammetry or laser scanning is often used to obtain accurate 3D models of existing structures. In the visualizing process, the decisions need to be made that will affect the final visual. This includes choosing the right materials and textures as well as making decisions concerning detailed modeling of particular monument segments. The challenge is to find balance between detail, computational requirements and available resources. Digital models and visualizations should be designed to be sustainable and easily updated as technology and scientific understanding continue to evolve. This will allow projects to adapt to new discoveries and technological advances. The current state of natural language processing and event extraction technologies presents challenges for the automated creation of historical narratives. Future work is focused on creating user-managed environments where users can contribute to event metadata, identify relations between events as well as create and share historical narratives (Akker et al. 2011, 5).

7.3 Artificial intelligence and 3D modeling

Artificial intelligence (AI) is a broad term that encompasses the development of computer systems capable of performing tasks typically requiring human intelligence. These tasks can include speech recognition, decision making, visual perception and even translation between languages. AI works on the principle of machine learning algorithms, which allow computers to learn from data and improve their performance without explicit programming for each individual task.

One of AI basic concepts is machine learning, where models gain the ability to predict or categorize data, which is based on previous experience. These models are gradually improving with the increasing amount of data they process. Recently, deep learning is widely used, which is a subcategory of machine learning that uses large neural networks with many layers to learn very complex data patterns.

The AI integration into the humanities is progressing more slowly compared to other areas, such as healthcare or business operations, but is already making significant contributions. These include improving the ability to process large amounts of data, increasing the efficiency of research processes and opening new possibilities for the interpretation and analysis of historical, literary and artistic works (Chapinal-Heras/Díaz-Sáncheza 2024, 1). AI is mainly used to process image data obtained by remote sensing, including satellite images and LiDAR (Light Detection and Ranging) technologies. These techniques allow rapid and accurate analysis of large archaeological sites, which significantly reduces the time and costs spent in conventional archaeological surveys (Puyol-Gruart 2002, Barceló 2006, 94–95). Models like ChatGPT can help automate routine tasks, such as writing messages or translating, and can also be a tool for creating illustrations or visualizations for archaeological purposes (Cobb 2023, 366–367). The use of AI in archaeology also includes advanced machine learning algorithms to process and analyze the data obtained. These algorithms can identify patterns and correlations that might be difficult to discern for human researchers. In this way, AI provides new analytical perspectives and depths that can contribute to new discoveries and better understanding of historical contexts (Chapinal-Heras/Díaz-Sánchez 2024, 4–5). In the field of archaeology, these tools can be expected to influence the research simultaneously with education and public presentations (Cobb 2023, 368).

In the field of monuments reconstruction and conservation, artificial intelligence can be important in many aspects. It can help analyze and document the condition of monuments using computer vision algorithms that can recognize and classify different building elements and materials. AI can also provide tools for 3D scanning and modeling as it allows experts to create detailed digital replicas of monuments, which is useful for accurate reconstruction or for virtual tours. During the reconstruction itself, AI can help identify the original materials and techniques used in construction by analyzing historical data, thus allowing more accurate and authentic restoration. Additionally, AI can optimize the reconstruction project logistics by predicting the amount of materials needed, estimating costs and timeframes, or monitoring the project progress (Diagram 5).

he AI also offers opportunities in the area of cultural heritage sustainability and conservation. For example, it can predict potential damage based on environmental factors or public access, thus enabling proactive conservation measures that can extend the monuments lifetime (Argyrou/Agapiou 2022, 2–3).

However, the use AI also requires consideration of ethical and practical issues, such as privacy protection, accountability and data accuracy. AI ethical principles should be clearly specified and balanced in relation to different contexts of use. This means that the principles should be formulated to consider specific situations and challenges associated with AI applications (Mittelstadt 2019). These ethical principles should be supported by a broad professional and public debate. This includes involving various participants in the process of formulating and revising the ethical principles to ensure that they reflect a wide range of views and needs (Mittelstadt 2019, 12). With the AI use, direct human interactions are reduced, which can lead to the creation of the so-called "moral distance". This distance can result in less ethical decisions because the decision-affected persons are not directly visible or are perceived only as abstract entities (Villegas Galaviz/Martin 2023, 1–2). It is important to take local contexts and cultural heritage into account when implementing AI technologies in the process of monument conservation in order to preserve the monument authenticity and historical value (Diagram 6).

Artificial intelligence (AI) opens up new possibilities in the monument and landscape reconstructions, as it offers effective methods for data collection,

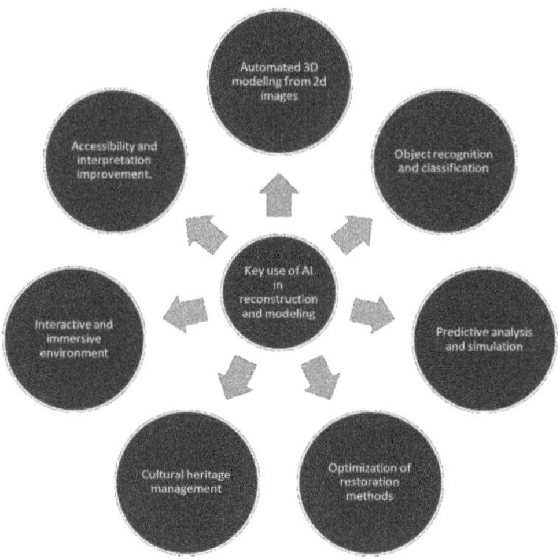

DIAGRAM 5

DIAGRAM 6

analysis, modeling and visualization. These technologies can significantly contribute to a more precise understanding and presentation of historical sites. Several essential uses of AI in this area are following.

Automated 3D modeling

AI can streamline the process of 3D model creating from 2D images or photogrammetric data by using machine learning algorithms. These systems are able to effectively recognize geometric shapes and textures, which enables faster and more accurate modeling of archaeological finds or architectural structures with no need for extensive manual intervention.

Object recognition and classification

Advances in deep learning allow for AI to be used to identify and classify historical artefacts from image data. This can simplify the process of sorting finds in archaeological excavations and help monitor the monuments condition and their protection as well.

Predictive analysis and simulations

AI can analyze large volumes of data about history and environmental factors to predict future changes that could affect monuments. These predictions can be used in planning the monument conservation measures or preparing for potential risks associated with climate change or urban development.

Optimization of restoration methods

AI can provide recommendations for choosing the most suitable techniques and materials for restoration of specific artefacts or structures based on the analysis of historical data and previous restoration projects.

Cultural Heritage Management

AI can efficiently process and analyze data from plentiful sources, including satellite imagery and digital archives, enabling better management, documentation and sustainability of cultural heritage. This involves managing digital repositories, optimizing searches and improving data sustainability.

Interactive and immersive environments

AI can provide visitors to museums and historic sites with personalized learning experiences through virtual reality applications that can adapt to previous user interactions and preferences.

Improving accessibility and interpretation

AI enables automated translations and delivery of content in different languages, which increases the accessibility and understanding of historical sites for international audiences. In addition, AI can help interpret complex archaeological data, providing visitors with deeper context and understanding of historical events.

Technological complexity of artificial intelligence (AI) represents a significant barrier to its wider adoption in the humanities. AI systems often require advanced technological skills and resources that may not always be available in the humanities. Here, this situation can relevantly limit the possibilities of using AI in standard research projects, where it could bring significant advantages in the area of data analysis and processing (Chapinal-Heras/Díaz-Sáncheza 2024, 4–5).

In addition to technological obstacles, the humanities often face the problem of a lack of specialized data. Humane disciplines are focused on historical and cultural data, which are often unstructured and very diverse. Acquiring and efficiently processing this data for AI purposes can be difficult and time-consuming. These data require specific methods for their digitization and analysis, which are outside the normal scope of machine learning algorithms used in other fields.

The use of AI in the humanities also brings ethical and methodological questions that cannot be ignored. There are concerns about privacy intrusions, the possibility for data misuse, and the risk of losing the human perspective in interpretation and analysis. These ethical challenges demand careful reflection and discussion among experts in order to have a balanced approach to the benefits and potential risks associated with the AI use in humane research (Chapinal-Heras/Díaz-Sánchez 2024, 4–5).

Artificial intelligence (AI) has the potential to dramatically improve the access and analytical capabilities of researchers in the humanities. Because of the AI ability to process huge amounts of data, researchers can provide deeper and faster analyses, while discovering new patterns and connections that might be missed by traditional methodologies (Puyol-Gruart 2002, 20). This increased ability to detect and interpret is essential, especially in cases where complex or historically/archaeologically extensive datasets are analyzed (Argyrou/Agapiou 2022, 3; Küçükdemirci/Sarris 2022, 6). AI also opens the door for interdisciplinary applications, allowing for combining methods and approaches from different disciplines. This merging of disciplines can lead to new perspectives and innovations in areas that have traditionally been separate. For example, integration of techniques from informatics and traditional historiography can support new ways of investigating historical events or literary styles (Puyol-Gruart 2002, 19–20; Chapinal-Heras/Díaz-Sánchez 2024, 4–5).

In addition to this, AI plays a key role in the cultural and historical heritage digitization and conservation. By using AI technologies to digitize documents, art and artefacts, these resources can be better protected and simultaneously more accessible for future generations. This process not only simplifies archiving and access, but also increases opportunities for global research and education in the field of cultural heritage.

Chapter 8

Uncertainty visualization and virtual modeling transparency

Visualization of uncertainty in the context of 3D modeling and reconstruction of monuments and landscapes is essential aspect for ensuring transparency and scientific integrity in interpretation of historical data or archaeological findings. This concept includes methods and techniques that allow the levels of uncertainty associated with reconstruction models and analyses to be explicitly displayed. Uncertainty can come from many sources, including data limitations, subjective interpretation, insufficient documentation, and methodological limitations.

Various methods can be used to visualize uncertainty. Colour codes and textures can indicate areas of the model with higher uncertainty, while less accurate or hypothetical parts can be shown with a different texture or greater transparency. Layers or animations are used to show various possible reconstructions, based on different interpretations or data, and allow visualizing how the reconstruction would change if specific elements are added or removed. In turn, interactive elements allow users to explore how different levels of confidence or uncertainty affect the overall model, giving them the ability to customize the model display according to different scenarios.

There is consensus in the scientific community that any level of uncertainty has a direct impact on our decision-making processes. For this reason, it is necessary to always formally express the accuracy and correctness of the data used, which enables a better understanding and communication of the results credibility.

If a specific data inaccuracy is known, it can be marked as an error. However, if the imprecision is not fully quantifiable, the term uncertainty is used, which involves a wide range of doubts and ambiguities (MacEachren et al. 2005, 140). In general, information uncertainty affects the process of analysis and interpretation, which can lead to biases against initial hypotheses, underestimating of negative evidence and overestimating of previous positive results (MacEachren et al. 2005, 140).

In order to express forms of ignorance in specific statements or hypotheses, it is convenient to categorize different types of nescience and associated uncertainties, even if a universal and unambiguous definition of this term still does not exist. Bonissone and Tong (1985) proposed three basic categories to express nescience:

1. Incompleteness – the value of individual variables is missing
2. Inaccuracy – the value of variable is known, but does not reach required accuracy level
3. Uncertainty – refers to situations in which the viewer creates their subjective impression because the information is not unambiguous.

Klir and Yuan (1995) relate uncertainty to lack of information and identify three basic types of uncertainty:

1. Non-specificity – when there are various interpretation options, but the information is imprecise or unambiguous
2. Indeterminacy – when the data are unclear and their meaning is vague or unlimited
3. Disharmony – when there is a conflict between possible alternatives and the data are contradictory.

These categories enable a better understanding of the nescience nature within theoretical considerations and modeling, while providing a framework for analyzing complex situations and interpretations.

Almost all data or pieces of knowledge can be accompanied by a certain degree of uncertainty, which can raise doubt in the viewer. If the information about this uncertainty is absent, the presented data can lost their key characteristics that describe how they were obtained, processed and interpreted. This information is essential for assessing the data quality and reliability before they are accepted and integrated into human knowledge (Zuk 2008, 2–3).

Techniques of information uncertainty visualization comprise plentiful approaches to help the viewer interpret the data presented. Systematic classification of uncertainty visualization methods involves, for example, adding of pictograms, modified geometry, attributes, animation and sound. Each visual stimulus is adapted to the data type, whether it is numeric, vector or categorical data. In the case of three-dimensional computer models, the following approaches can be applied (Griethe/Schumann, 2006):

- Free graphic variables – can affect colour shade, texture, lighting, sharpness, clarity; sequence of colours can be used as well
- Comparing of visualizations ("side by side") – comparing of two or more visualizations side by side allows for identifying differences among various interpretations
- Integration of objects – adding of other objects, such as pictograms, graphs, tables or texts can provide additional information and context.

MacEachren and his team (2005, 155–157) identified seven key challenges related to uncertainty presentation and disclosure of in geographic data. These tasks require interdisciplinary collaboration, as working with uncertainty involves both technical and methodological aspects (for example, how to quantify and display the uncertainty exactly) and understanding of psychological, social and contextual factors that influence the way how uncertain information is perceived and interpreted by people. The goal is to develop tools, methodologies, and theoretical frameworks that will help professionals and users work with uncertainty more effectively and consider it in analysis and decision-making. The key challenges are following:

1. Understanding the individual components of uncertainty and their relation to the area of interest, users and information needs.

   Research on uncertainty across disciplines and practice areas is still limited. Most works are focused on specific uncertainty types, such as uncertainty in spatial or temporal data; less attention is paid to abstract forms of uncertainty, such as data completeness or consistency.

2. Understanding how knowledge of uncertainty affects analysis, decision-making and outcomes.

   Despite extensive research on uncertainty visualization, few studies have been devoted to its impact on the decision-making process. It is important to understand how the uncertainty affects decision-making and to create formal models that allow the uncertainty to be effectively factored into various scenarios.
3. Understanding how uncertainty visualization can help the research analysis.

   Visualization is often used for data analysis, but the question remains how the presence of explicitly visualized uncertainty affects the data exploration process and whether data interpretation is more effective if uncertainty levels are directly displayed.
4. Elaboration of methods for capturing uncertainty in individuals, who analyze the data and make decisions.

   If uncertainty is associated with the measurement of certain parameters, it can be expressed numerically. However, if uncertainty is connected to human judgment, it is very difficult to capture and express it in a structured way when processing the information.
5. Developing of methodologies for visualization of various uncertainty types.

   Despite the existence of various approaches and techniques for uncertainty visualization, there is no comprehensive manual or set of rules that would clearly define how different uncertainty types can be visualized in different contexts.
6. Developing of tools for interactive uncertainty visualization.

   Interactivity as an essential component of visual environments still does not receive enough attention in research on uncertainty visualization. A comprehensive framework for understanding how interactive interfaces can support manipulation of uncertainty visualization and facilitate users' dynamic data exploration is still missing.
7. Assessing of applicability and usefulness of methods aimed in uncertainty capturing and presentation.

   New empirical methods need to be developed to assess the effectiveness of visual strategies in solving vaguely structured problems. Assessment of appropriate approaches should factor their effectiveness in communicating uncertainty and facilitating decision-making processes.

In order to achieve graphical clarity, exactness and efficiency of visualizations, the guidelines have to be followed that minimize displayed data distortion, support data comparison, allow displaying a large amount of information in a small space and reveal multiple levels of data details. It is also important to closely link statistical and textual descriptions. For this reason, several basic principles need to be observed:

- Proportionality of graphic presentation: Graphic representation of numerical data should be directly proportional to the data displayed.
- Clarity of text and labels: Text and descriptions have to be understandable and unambiguous to avoid ambiguities.
- Display of variations of data, not design: Visualization should emphasize variations in the data, not in the design itself.
- Adequacy of the number of dimensions: The number of dimensions used to display the data should not exceed the number of dimensions of the data presented.
- Contextualization of the data: Data should never be presented out of context.

Another important principle is "data-ink maximization", which encourages visualization creators to present as much data as possible with the minimum consumption of graphic elements. Data-ink refers to graphic elements, removal of which would violate the integrity of the displayed data. All graphic elements that do not represent the data should therefore be eliminated (Tufte 2001; Zuk/Carpendale 2006).

Main factors influencing contextual and spatial uncertainty can be caused by the type of data available, their completeness, reliability and interpretation. In visualizations, these issues can be related to location geometry, age, colour, texture, material, construction, context, and landscape (Brusaporci 2017, 142).

In the reconstructed model, three areas are affected by uncertainty:

1. Shape - includes the object geometry, size and spatial location
2. Material - refers to the building physical form and stratification
3. Appearance - is related to surface features such as texture and colour (Appollonio 2016, 177).

To avoid these uncertainties and ensure transparency and credibility of digital visualizations of cultural heritage, the London Charter was introduced in 2009. This charter provides guidelines for setting standards for creating and presenting the digital models so that they are reproducible and reliable. The London Charter emphasizes documentation and methodology, thus ensuring that the created digital models are supported by relevant data and research (Denrad et al. 2012; www.londoncharter.org).

In Spanish-speaking countries and Latin America, the Sevilla Charter is aimed to extend and apply the London Charter. It represents adaptation of these principles to local needs and contexts, emphasizing cross-cultural exchange and collaboration in the field of cultural heritage digital visualization. The Sevilla Charter also emphasizes education and spreading of approved practices in this area (Karlsson 2013, 47; Lopez-Menchero/Grande 2011).

Compared to physical models, digital models have the advantage of less time and more flexibility in editing by specialized software. However, the rise of new technologies has brought both benefits and many issues and conflicts, among which the input data transparency and accuracy is one of the most discussed (Gershon 1998, 43).

Based on the questionnaire, Maria Sifniotis and the team of authors (2007, 2) created a reliability table of various sources, in which written sources are considered to be the least reliable and preserved objects are the most reliable.

Foundations of cultural heritage digital transformation in European Union

In 2000, a meeting was held in Feira, Portugal, aimed at creating an eEurope action plan, the aim of which was to fully exploit the potential of digital technologies. There, a coordination mechanism for the cultural heritage digitization in EU member states was created. The importance of the cultural heritage digitization was emphasized again in 2001 at the meeting of EU experts in Lund, Sweden, where the member states committed to invest in programs supporting the digitization of museum collections, archaeological sites, archives, maps and historical documents (Lopez-Menchero/Grande 2011).

Growing concerns about the uncontrolled development of computer visualizations led to the founding of the Virtual Reality Culture Organization (VRCO) in 2000. In the same year, the Kings Visualization Lab launched the "Making Space" project, which was aimed in creating a methodology for documenting the three-dimensional visualization cognitive process. In 2006, fifty outstanding experts discussed the issues of computer models transparency at the British Academy. The meeting resulted in publishing a document that, under the leadership of Franco Niccoluccio, laid the foundations of the London Charter.

The impact of virtual reconstructions on viewers, compared to schematic drawings, was criticized in 1999 already. At that time, the recommendation was to reduce the level of detail, abandon photorealistic visualizations and to concentrate on expressing the degree of uncertainty and interpretation of the resulting model (Strothotte/Masuch/Isenberg 1999, 36–37). The solution to several problems in the field of virtual reconstructions could be incorporation of additional data, which would allow monitoring the decision-making processes related to the uncertainty in data and creating of reconstruction proposals. Instead of formulating uncertainty in words, providing the computer with quantitative data is important. A possible solution would be to create a classification based on coded information that would indicate the uncertainty degree.

Visualization of virtual models transparency can be realized in different ways, for example by using colours and transparency, choosing the type of rendering and surfaces, or adding text and multimedia information that can be linked to the data via the Internet (Kensek/Dodd/Cipolla 2004, 175–180). The key element is a detailed description of each part of the modeled structure, its spatial and temporal transformations, as well as documentation of the decisions that were made during the visualization designing. In this way, the documentation will be created that captures the uncertainty level of individual data. All this information should be organized in a database structured correspondingly to the model. The visualization output can include a graphical uncertainty representation in the form of shading, transparency, colour shades, simplified textures, geometry, information labels or multiple windows (Brusaporci 2017, 82).

The process of working with digital data is detailed in several manuals, such as 3D-ICONS (2014) or IT-Empfehlungen für den nachhaltigen Umgang mit digitalen Daten in den Altertumswissenschaften (2017). However, these methodologies mostly deal with digital documentation, modeling or presentation, but do not provide solutions to the problem of data uncertainty visualization.

An important aspect is the working procedure documentation and the distinction between the parts of the model that are reconstructed on the basis of real data and those that are completed hypothetically. Each part of the reconstructed model should contain information about the uncertainty of the data on the basis of which it was created.

Understanding of the data uncertainty is perceived differently in various projects. The cultural heritage visualization should be consistent with the accuracy of traditional research, where this aspect is clearly evident. Any high-quality visualization should precisely inform about the differences between real data and hypotheses, as well as about different levels of probability (Denard 2012, 60).

Most uncertain data in computer reconstructions are visually marked using a colour scale, combination of transparency and colour shades or varied forms of displaying the model surface, such as wireframe or different types of textures. Other methods involve darkening or lightening particular elements, or changing the thickness of lines to express varying credibility degrees. These methods solve similar problems, such as representing multiple uncertainty types simultaneously or combining realistic surface with a probability degree.

In most cases, uncertainty visualization should clearly and effectively show the differences between various data types and express the levels of certainty and precision. These approaches not only improve the viewers′ awareness, but also promote credibility of digital reconstruction results, thereby contributing to their greater transparency and scientific authenticity.

Reconstruction creating procedure

The process of creating a reconstruction has to be defined very accurately, as it is necessary for verifying the possible results, reusing of the information and also for better interpretation of the digital model created. Appollonio (2015) suggests structuring the reconstruction process as follows:

1. Collection of documentary sources

   Available materials, such as plans, photographs, written sources and other sources that can provide relevant information for the reconstruction have to be collected systematically.

2. Semantic structuring of artefacts
   All the information obtained has to be classified and organized into semantic categories to make further data processing and interpretation easier.
3. Analysis of documentary sources and extrapolation of information about the artefact invariable properties
   The analysis, interpretation and decision-making has to be based on the collected evidence. The analyzed properties involve geometric shape, surface appearance and physical properties of artefacts. This step includes deduction, extrapolation and possible conjectures.
4. Correlation between the data used in the reconstruction process and their degree of uncertainty
   The uncertainty degree of each individual element of the model has to be determined based on available data.
5. Reconstructive 3D modeling
   The three-dimensional model has to be reconstructed based on analyzed and interpreted data.
6. Semantic expansion of the 3D reconstructed model
   The 3D model has to be supplemented with additional semantic information that increases its complexity and informativeness.
7. Validation and reconstruction hypotheses
   Hypotheses obtained on the basis of expanded data for each basic or essential element of the model and their graphical or textual representation have to be verified.

Categorization of source data uncertainty

Appollonio (2015) suggests categorizing source data uncertainty according to the accuracy of geometric, spatial, formal, material and construction properties that can be obtained or derived from available sources. The uncertainty degrees are defined as follows:

1. Reconstruction based on archaeological or architectural evidence
   (real data, graphic documentation of losses)
2. Reconstruction based on original drawing documentation
   (lower level of accuracy of spatial data)

3. Reconstruction based on construction data referring to the same period analogies
   (e. g. covering, type of roof, frame construction, basement solution, windows, doors, decorations)
4. References to literature directly from the author of reconstruction
5. References to literature related to the studied issue
6. Interpretive hypotheses based on the study of architectural style and time period
7. Interpretive hypotheses from other research areas related to the construction systems of the given historical period (can provide structurally plausible solutions)
8. Reconstruction hypotheses without references.

Uncertainty processing and transfer

According to Schäfer (2018, 91), it is important to focus on the following key stages for the uncertainty transfer and spreading in processing the background information, its use and resulting interpretation:

- Data acquisition and collection – Classification and categorization of data
- Data processing – Filtering, transformation, sampling, quantification, simplification, etc.
- Visualization – Mapping, modeling, rendering
- Evaluation – Human factor and incomplete knowledge
- Presentation and discrimination – Data form, operating system, decontextualization.

Since uncertainty is present in all steps of the virtual model creating process, it is important to carefully document this process, including all ambiguities, problems and decisions made. Transparency of information about the entire cognitive process is essential to maintain the virtual reconstruction credibility. Subjective decisions and assumptions, which can significantly affect the uncertainty degree, open up possibilities for alternative reconstruction and are often not declared (Appollonio/Giovannini 2015, 7), have to be clearly documented as well.

Reliability index

For each reconstructed part of the model, reliability index can be used, which expresses the author's certainty or, on the contrary, doubts about the correctness of the classification of a part of the model. This index may reflect the author's confidence in the accuracy of his interpretation, based on primary documents, analogies, comparative research, and scientific deduction (Hermon/Nikodem 2008, 2). Another criterion for evaluating the model credibility can be the importance index, which reflects the importance of a particular part for the whole model and its connection to the goals that the visualization is intended to meet.

Uncertainty visualization

Visual presentation of three-dimensional virtual data offers four basic approaches to display data uncertainty: wire (mesh) mode, mass mode, texture mode and material mode of working with the model. Each of these formats allows encoding the necessary information in different ways. However, the basic method of visualizing uncertain data is still the use of colour scales, which can represent various values, attributes and other information. These visualization options offer a wide spectrum of combinable display methods. For better illustration, the model can be rendered in wire (mesh) mode, for example, which can help to better understand complex structure of the object or details of building constructions (Reilly 1992, 156; Baker/Meulenberg/Rode 2003, 2). This method can also be used to express uncertainty. The parts of the model that are based on real information can be colour-coded, while wireframe mode can be applied to the parts where the uncertainty index falls below a predetermined threshold.

The form of uncertainty visualization is usually determined by the rules of graphic design. For reconstructed models, this visualization can be significantly altered, as it is often associated with data interpretation. For example, using a photo-realistic style may give the impression of high certainty, as the object geometry or appearance is reconstructed based on precisely measured data. Photorealistic models usually contain such a high level of detail that their ability to display uncertainty is the subject of long debates, which usually conclude that they are not suitable for academic setting (Olivito/Taccola 2004, 181; Wittur 2013, 48). The trend of displaying virtual models and scenes in a photo-realistic way has emerged in the beginning of virtual reconstructions and still continues. Critics of this style agree in particular that photorealistic rendering does not distinguish between real and interpreted structures (Kantner 2000, 47). However, there are also opposing approaches arguing that, if used wisely, photorealistic rendering can significantly help in visualizing the cultural heritage, as the visualization technique should depend on the audience targeted (Roussou/Drettakis 2003).

Considering the rules based on international conventions, it can be concluded that the photorealistic style is not suitable for uncertainty displaying and should be excluded from scientific reconstructions. In contrast, simple abstract styles are often used to depict uncertain elements, where a simple geometric shape and reduced number of details are used, while increasing the level of certainty of the entire object. Schematic rendering can thus emphasize uncertain parts and indicate the speculative character of the displayed objects, as this principle is also commonly used in architectural visualizations (Strothotte/Masuch/Isenberg 1999, 36). In applying this rendering style, details are almost completely excluded to highlight the data uncertainty. It can also be advantageous to combine photorealistic elements with schematic ones that represent the interpreted data. This approach uses data from three-dimensional documentation, where a model based on real data is combined with simple geometric shapes indicating non-existent structures. In this way, photorealistic parts with a high degree of detail and certainty can be distinguished from abstract modeled parts.

Textures are essential element of photorealistic practices, which can contain images of real objects and, as a result, display colour and material with high exactness. Textures can be created directly from objects in the field (in situ) and their use for reconstructed parts can bring a high degree of certainty. Final uncertainty can be represented by symbols, such as simple lines separating individual uncertainty degrees, which can be assigned to uncertainty attributes using different colours or line thicknesses (Hermon/Nikodem 2008, 39; Strothotte/Masuch/Isenberg 1999, 39). Another way of uncertainty visualizing can be creating of multiple models to compare different hypotheses. The advantage of this approach is that it is sometimes difficult to discuss the construction details without proper consideration in the wider context. The disadvantage is that it is a time-consuming process, since each variant have to be modeled from scratch. The need of creating a new model for each of the theories can be partially solved with the help of interactive browsing, which allows displaying of various versions of individual structural elements.

In a practical context, integrating of uncertainty visualization into all phases of a 3D monument modeling project is essential. During the planning phase, it is important to clearly define where and why the uncertainty may arise and to design methods to visualize it. In the implementation phase, the uncertainty visualization techniques need to be incorporated directly into the model creation process so that all project participants are informed about the possible model limitations. When presenting the results, the places where and how the uncertainty was visualized and what impact it has on the interpretation of the model have to be communicated clearly. This transparent display and communication of uncertainty can help ensure that all those involved in the project – from the research team to the public and professional community – have a proper understanding of the 3D monument model limits and possibilities.

Chapter 9

Evaluation of reconstruction relevancy

Uncertainty can occur in many forms and for different reasons, and its correct identification and classification are crucial for effective decision-making and planning. In the context of 3D modeling and monument reconstruction, uncertainty classification helps distinguish between different types and sources of uncertainty, allowing a better understanding of potential risks and limitations associated with model interpretation and analysis. Classification of uncertainty is important for several reasons, firstly because it provides a better understanding of the uncertainty nature and causes, enables more accurate risk management, and improves the quality of decision-making in various fields, including scientific research.

One of the main reasons why classifying uncertainty is useful is that it allows for a clearer understanding of its nature and sources. This is critically important as various types of uncertainty require different approaches to manage and minimize them. For example, epistemic uncertainty as a consequence of lack of knowledge can often be mitigated by further research and data collection, while random uncertainty that is inherently unpredictable may require strategic planning and provisioning –reserve plan. The correct uncertainty classification enables better risk management as well. When uncertainties are properly identified and categorized, targeted measures can be taken to mitigate their impact. For example, in engineering projects where uncertainties can have significant financial and safety implications, managing uncertainty can be fundamental to the development of robust and safe systems. Uncertainty classification also significantly contributes to the improvement of decision-making processes. When these processes are informed about different types of uncertainties, decisions can be made with greater exactness. In the context of environmental policy, for example, distinguishing of various types of uncertainties can help develop policies that can better consider potential risks and uncertainties associated with climate change.

Finally, effective classification of uncertainty increases transparency of scientific and technical communication. When organizations and researchers are open about the uncertainties associated with their work, it can increase confidence in their conclusions and recommendations. Transparency on issues of uncertainty can also endorse professional debate and critical thinking, which can lead to improved scientific and technical methodologies.

Our uncertainty classification system designed for 3D reconstructed models of immovable architectural and historical monuments is based on the model proposed by J. Unger (2019) in his dissertation work (Table 2). According to J. Unger (Tab. 2, Unger 2019, 75), in the context of archaeological objects reconstruction, the uncertainty classification is crucial in distinguishing various levels of exactness and credibility of information obtained from archaeological finds and their interpretations. This approach allows the scientific community and experts to better understand the limits and potential range of interpretations that can be based on the available data.

Unger's classification methods are based on several elemental principles that ensure systematic and objective approach to the uncertainty assessment:

1. **Transparency:** Each step in the classification process should be clearly documented and transparent in order to be understandable to other researchers and to enable the results reproducibility.
2. **Evaluation of uncertainty sources:** Identification and analysis of possible uncertainty sources, including data errors, subjective interpretations and methodological limitations. This process involves distinguishing between epistemic uncertainty (lack of knowledge) and aleatoric uncertainty (random variability).

Table 2 Uncertainty classification of archaeological monuments according to J. Unger. Created by Unger 2019, 75; Edited by D. Bešina.

	100	50	0
Preserved	whole/most part	remains/extensive reconstruction	none
Space localization	located/locatable	old plans & maps / other sources	estimated
Archaeological/HMM research	yes	partial	none
3D digital documentation	yes/most part	partial	none
Detailed location / object building plans	yes	partial	none
Iconographic sources	3 and more/ or Willenberg	1 or 2	none
Clear analogy	yes	partial	none
Written sources	detailed	mentions	none
Texture	real	derived	no context
Architectonic context logic	yes	partial	none

HMM – Historic Monument Management

3. **Categorization:** Uncertainties are categorized according to their causes and nature, thus allowing the scientific community to better comprehend how different types of uncertainty can affect conclusions drawn from archaeological data.
4. **Uncertainty visualization:** Use of graphical and visual tools to display uncertainties, which may include colour coding, textures or layering on maps and 3D models. These techniques help intuitively present uncertainty levels associated with different parts of the model or map.
5. **Iterative approach:** Uncertainty classification should be considered an iterative process where results are regularly revised and updated based on new data or a better understanding of uncertainty sources.

In practice, J. Unger's (2019) uncertainty classification is applied in modeling and reconstruction of archaeological objects in such a way that each phase of the process is analyzed and evaluated in detail. This allows researchers and archaeologists to consider various interpretations and scenarios that may arise from the analyzed data. This makes it possible to approach the reconstruction of historical events, cultural practices or ancient technologies more objectively, which helps to create more exact and veracious representations of the past. This approach also provides a robust platform for discussion and critical analysis of scientific findings, which is essential for academic rigor and ethics in archaeological research.

The uncertainty classification method, as proposed by J. Unger (2019), can be used as a basis in designing the classification of immovable architectural-historical monuments. This approach allows dealing with uncertainties and various levels of available information systematically, which can significantly help in architectural monuments protection, conservation and management. In the classification process, transparency and complete documentation of each step has to be accented. When classifying architectural monuments, this would involve recording of all available historical and current data about the object, including architectural plans, historical photographs, written records and data obtained by using modern technologies such as 3D scanning. This documentation should be accessible to all who are involved, including the scientific community, monument managers and the public.

The classification should identify which aspects of the monument's condition or history are least certain and why. This could include questions regarding its original use, the materials used or the dates of construction phases. Various methods, such as dendrochronology (dating by tree rings) can help reduce uncertainty in some ways, while in others it may be more natural, for example due to the lack of written records. Visualization of the uncertainty level within information panels or digital models of monuments could be realized by various tools. Colour codes, different textures or stripes can indicate areas where information is incomplete or where divers interpretations exist. These visualizations could be part of interactive displays in museums or online databases, thus allowing visitors and researchers better understand where further research is needed.

The uncertainty classification should be considered a dynamic process, with regular updates as new information becomes available or as technologies for obtaining it improve. This approach will ensure that the classification reflects the latest knowledge and is as precise as possible.

The uncertainty classification of 3D reconstructed models of immovable architectural-historical monuments suggested by us represents a comprehensive and adapted approach, which is based on Unger's model (2019) and the model proposed by Apollonio (2015), but modified to meet specific needs of architectural monument context in a better way. This system uses a multidimensional evaluation of various aspects that can influence the accuracy and authenticity of reconstructed 3D models.

This classification introduces a multidimensional scale that evaluates every aspect of monument reconstruction within the scope from complete authenticity to complete uncertainty. This system includes criteria such as building conservation, degree of building-historical research, level of building detailed localization, application of BIM (Building Information Modeling), availability of 3D digital documentation, existence of historical technical plans, vedutes, historical drawings and photographs, as well as access to written sources, textures and materials. Compared to the original Unger´s system, which may have been more general and less specific to architectural monuments, this classification adds specific criteria relevant to the architectural sphere. These criteria include direct analogy, which is important when comparing similar existing objects, and spatial context in situ, which determines if the monument is in its original environment. These elements are crucial to understanding how a monument fits into the wider historical and cultural context.

This classification system brings significant benefits to practice, especially in the areas of research, conservation, restoration and presentation of immovable monuments:

1. **Increased accuracy and transparency:** Allows more accurate identifying and communicating of uncertainty levels associated with various aspects of reconstruction works. This is critical to ensuring scientific rigor and ethical integrity in the monument care.

2. **Better management and planning of conservation interventions:** Provides tools for more effective planning of interventions that are sensitive to historical and material context.
3. **Support for education and public awareness:** Using this system, educational programs and materials can be created to illustrate how scientific knowledge about monuments evolves and uncertainties are gradually overcome.
4. **Increasing international cooperation:** Standardized approach to assessing and classifying uncertainty can facilitate international cooperation and sharing of best practices in the field of monument care.

The uncertainty classification table below (Tab. 3.) for 3D reconstructed models of immovable architectural-historical monuments is designed to

Table 3 Uncertainty classification of architectural-historical immovable objects. Created by D. Bešina.

Preserved	100 whole/most part	50 remains/extensive reconstruction	0 none
Building-historical or archaeological research	yes	partial	none
Detailed location /object building plans	yes/most part	partial	none
BIM realization	yes/most part	partial	none
3D digital documentation	yes/most part	partial	none
Historical technical plans and failures	2 and more	1	none
Vedutes or historical drawings	3 and more	2 or 1	none
Historical photographs	2 and more	1	none
Direct analogy	yes	partial	none
Written sources to the monument shape	yes	partial	none
Texture	real	derived	no context
Material	real	derived	no context
Spatial context in situ	original	changed	none

enable comprehensive and forced evaluation of various aspects related to information accuracy and authenticity in reconstruction models. The table contains a series of criteria, which are divided into columns according to degree of uncertainty, where each criterion is assigned a score from 0 (greatest uncertainty) to 100 (least uncertainty).

Principles of classification

Evaluation of criteria

Each criterion is evaluated according to three possible levels that reflect different degrees of uncertainty:

- **Yes/for the most part (100 points)**
- **Partial (50 points)**
- **None (0 points)**

This three-level evaluation allows considering various levels of information availability and accuracy for every aspect of the monument.

Evaluation criteria

1. **Building preservation:** The extent, to which the original building is preserved, is evaluated.
2. **Building-historical or archaeological research:** Considers if there was a research realized on the site that can contribute to the reconstruction exactness.
3. **Detailed localization of object:** The level of details, which were obtained and documented, is assessed.
4. **Realization of BIM (Building Information Modeling):** The use of modern BIM methods in documentation is determined.
5. **3D digital documentation:** The availability of the object 3D documentation is examined.
6. **Historical technical plans and failures:** The number of available period plans and failures is evaluated.

7. **Vedutes or historical drawings:** The number of available views or drawings that help illustrate the object appearance in the past is quoted.
8. **Historical photographs:** The number of available historical photographs of the object is quoted.
9. **Direct analogy:** The existence of similar objects that can serve as a reconstruction reference is stated.
10. **Written sources regarding the object shape:** The written sources about the object available are examined.
11. **Texture and material:** Evaluates if real or derived materials and textures are used in 3D model.
12. **Spatial context in situ:** Assesses if the object is in its original environment or has been moved or changed.

This classification brings significant assets to practice by providing a clear and structured framework for evaluating uncertainty in the reconstruction of immovable monuments. This approach allows for better understanding the potential risks and limitations associated with available data and interpretations. The table can also support transparency in the decision-making process and ensure that all relevant information is considered in the process of planning the monument conservation, restoration or its presentation.

Uncertainty classification can be used in virtual reconstructions of historical landscapes to improve accuracy, reliability and interpretation of results. This approach allows for the uncertainties associated with reconstruction to be systematically identified and quantified, thus increasing confidence in final models and visualizations. Virtual reconstruction should include assessment of uncertainties associated with each geographic and historical data. This may involve geological, archaeological, architectural and historical information. Each piece of information should be categorized according to uncertainty degree, which can be determined by evidence availability and consistency. For example, data obtained from well-documented archaeological excavations could be considered more reliable than those derived from less investigated sources.

For this purpose, we have designed our own uncertainty classification system for 3D reconstructions of historical landscapes in 3D virtual space (Table 4).

Our uncertainty classification table for 3D reconstructions of selected historical landscapes in 3D virtual space is designed to provide clear and structured framework for evaluating various aspects that can affect reconstruction

Table 4 Uncertainty classification for 3D historical landscape reconstruction in virtual 3D space. Created by D. Bešina.

	100	50	0
Changes in geo-relief	none	partial	considerable
Changes in micro-relief	none	partial	considerable
Changes of original landscape cover	none/most part	partial	considerable
Research on anthropogenic phenomena	yes	partial	none
Research on environment	yes	partial	none
Anthropogenic structures	preserved/most part	partial	none
Historical mapping – small scale	3 and more	2 or 1	none
Historical mapping – medium scale	3 and more	2 or 1	none
Historical mapping – large scale	3 and more	2 or 1	none
Historical aerial photographs	yes/most part	partial	none
Detailed cartographic localization (LiDAR, DTM, DMR)	yes	partial	none
3D digital documentation (drone)	yes/most part	partial	none
LIM realization	yes/most part	partial	none
Vedutes or historical drawings	3 and more	2 or 1	none
Historical photographs	2 or more	1	none
Written sources	detailed	mentions	none

accuracy and authenticity. This classification system enables a systematic and objective evaluation of data and information according to precisely defined criteria that are focused on various characteristics and properties of historical landscape.

Again, the classification system uses a point rating divided into three categories (100, 50, 0 points) based on information accuracy and availability. Each criterion, such as changes in the area's geo-relief, micro-relief, original landscape cover, anthropogenic structures, and historical mapping, is evaluated and scored depending on the extent to which information is available and veracious. For example, the complete preservation of anthropogenic structures and the availability of multiple historical maps at small and large scales will yield a higher score, while incomplete or absent data will receive a lower score.

Unlike other classification systems that may provide more general or less structured methodology, the proposed classification system allows for an accurate and specific assessment of each aspect of the historic landscape. This approach does not only consider the data availability, but also their historical conceptual value and contribution to the reconstruction veracity. The system improves the accuracy of reconstructions by providing a clear and quantifiable assessment of uncertainty. This allows experts and historical geographers to better understand which reconstruction parts are most evidence-supported and where potential weaknesses or speculation are. By providing a clear scoring system for evaluating various landscape aspects, this system allows for more efficient decisions about where further research and resources have to be focused. This is particularly useful when allocating limited resources, such as deciding which areas of a historic landscape need further archaeological or geological investigation.

Determining the value of each criterion in uncertainty classification is crucial for more precise evaluation of the accuracy and reliability of historical landscapes 3D reconstructions. The evaluation criteria are following:

1. **Changes of the territory geo-relief:** This criterion evaluates changes in geo-relief of the given territory within the scope of none, only partial or complete.

2. **Changes in the territory micro-relief:** Changes in the micro-relief of the territory are evaluated, with the possibilities of evaluating none, only partial or complete changes.
3. **Changes of the original landscape cover:** This criterion deals with changes in the original landscape cover, where complete preservation, partial changes or significant changes are evaluated.
4. **Research on anthropogenic phenomena in the territory:** The level of archaeological and anthropological analysis carried out in the given territory is evaluated within the scope from complete analysis to partial to no research realized.
5. **Research on the territory environment:** Research of environmental factors in the given area is evaluated, with the possibilities of evaluating complete, only partial and no research.
6. **Anthropogenic structures:** This criterion evaluates preserved anthropogenic structures in the territory, with the possibilities of evaluating full preservation, only partial preservation and complete absence.
7. **Small-scale historical mapping of the territory** (from 1:1000 to 1:5000 - simplified content, no contour lines, plans): Availability of historical maps on a small scale is introduced, with the possibilities of 3 and more, 2 or 1, and no historical mapping.
8. **Medium-scale historical mapping of the territory** (from 1:10,000 to 1:200,000): Availability of historical maps on a medium scale is introduced, with the possibilities of 3 or more, 2 or 1, and no historical mapping.
9. **Large-scale historical mapping of the territory** (from 1: 250,000 to 1: 1,000,000): Similar to the previous criterion, availability of historical maps on a large scale is evaluated.
10. **Historical aerial photographs:** Availability of historical aerial photographs is evaluated, with the possibilities of yes/for the most part, partly or none.
11. **Detailed cartographic localization of the territory** (LiDAR, DTM, DMR): Availability of detailed cartographic images of the territory is evaluated, with the possibilities of yes, partial and no localization.
12. **3D digital documentation (drone):** The criterion evaluates available 3D digital documentation created by using drones, with the possibilities of yes/mostly, partially and no such documentation available.

13. **Implementation of LIM:** The criterion evaluates whether landscape information modeling (LIM) was implemented for the territory, with the possibilities of yes/for the most part, partially and no.
14. **Vedutes or historical drawings:** This criterion evaluates availability of vedutes or period drawings of the area, with the possibilities of 3 or more, 2 or 1, and no views or drawings available.
15. **Historical photos:** Availability of period photos of the area is evaluated, with the possibilities of 2 and more, 1 and no photos available.
16. **Written sources:** Availability of written sources referencing to the area is evaluating within the range of detailed information, only mentions or no written sources available.

All the classification criteria proposed by us for evaluating the relevance of reconstructions dependent on the uncertainty visualization can serve as a starting point for evaluation of different contexts, while the point evaluation reflects their importance or quality. In practice, both tables can be adapted according to specific needs and criteria that have to be taken into account when evaluating the contexts of monuments. It is important to realize that, in both cases, mathematical evaluation can be subjective and dependant on the evaluator's or research team's preferences and expectations. Therefore, the tables can serve as a tool to guide the assessment and compare contexts, but the final evaluation should consider qualitative factors and the wider context of the monuments.

The context of monuments and landscapes can undergo changes for various reasons. Anthropogenic changes, such as urbanization, industrialization, or changes in agricultural or forestry activities, can dramatically alter the surrounding context of monuments and landscapes. Natural events, such as floods, landslides, or forest fires can impact on the landscape significantly and affect the monuments situated there. In addition to this, climate changes can affect vegetation, watercourses and other landscape features, which can influence monuments neighbourhood and their visibility. Air, soil and water pollution as well as landscape degradation can cause damage to monuments and their surroundings. New archaeological discoveries can change our understanding of the history and context of monuments, which can determine the way they are presented and interpreted. Changes in social values, traditions and preferences can also affect the way monuments and their landscape are perceived and interpreted. In the context of historical monuments and cultural landscapes, these changes are inevitable part of historical development dynamics. It is important to monitor them and consequently adjust the conservation and research measures to preserve the cultural heritage and present it in the context of its current environment.

For this reason, it is useful to have your own evaluation tool for changing interpreted contexts. We present a proposal for the evaluation of historical and landscape contexts, which can be included in the overall calculation result of the reconstruction relevance (Table 5). The evaluation in question does not primarily need to be included in 3D reconstructions of objects and environments, as it is in the case of the tables above, but it can rather be used in identifying, protecting and designing a presentation in situ.

This classification table provides a system for evaluating monuments and their contexts from multiple perspectives, using a point rating from 0 to 5, with zero meaning no contexts. Each criterion in the table considers different aspects that affect the historical, architectural, landscape, social, cultural and economic significance of the monument.

Historical context: These criteria evaluate significant events, chronological context and historical relevance of a monument. Evaluating these factors can help determine how important the site is from the historical point of view and how it is connected with events and history of the region.

Architectural context: Criteria in this category evaluate the uniqueness of style and techniques, the materials and technical details used as well as the monument proportions and aesthetic value. These factors are important when assessing the architectural significance of the monument and its place in the context of building styles and techniques of the given period.

Landscape context: The way how the monument is integrated into its natural environment, its position in the landscape, accessibility and views as well as its relationship to nearby anthropogenic and architectural structures allows evaluating the monument significance within the surrounding landscape and its place in the landscape context.

Social and cultural context: These criteria evaluate the level of residents' participation, importance to the local community and cultural and religious customs. These factors help determine social and cultural

Table 5 Evaluation of a monument and landscape mutual context. Created by D. Bešina.

Context	Criteria	100	50	0
Historical	Important events			
	Chronological context			
	Historical relevancy			
Architectonic	Unique style and techniques			
	Materials and technical details			
	Proportions and aesthetical value			
	Connection with the residence urbanism			
Landscape	Integration into natural environment			
	Important site in the landscape			
	Access and sightseeing			
	Direct link with other anthropogenic structures in the vicinity			
	Direct link with other architectonic objects in the landscape			
Social and cultural	Residents participation			
	Importance for local community			
	Cultural and religious associations			
Economic	Contribution to local economy	relevant		
	Potential for tourism			
	Investments in protection and restoration			
TOTAL				

significance of the monument and its place in the collective memory and identity of local residents.

Economic context: The criteria of the monument´s contribution to local economy, its potential in terms of tourism and investments in the cultural heritage protection and restoration can help evaluate economic significance of the monument and its potential to contribute to economic development of the region.

In order to evaluate the overall result of the uncertainty classification for the evaluation of the reconstruction of monuments in the landscape context, expressed as a percentage, a scoring according to the tables need to be calculated. To calculate the percentage result from the total score, maximum possible score that can be achieved has to be determined first, and then divide the total score by the maximum score and multiply by 100.

Example:

First, the maximum possible score for each table is determined. They assume that each criterion in each table can reach a maximum of 100 points. Based on the number of criteria in each table, they calculate the maximum possible score for each table.

- For Table 1: Monuments have the number of criteria 13, so the maximum possible score is 13 * 100 = 1300 points.
- For Table 2: Historical landscape has the number of criteria 15, so the maximum possible score is 16 * 100 = 1600 points.
- For Table 3: Context has the number of criteria 18, so the maximum possible score is 18 * 100 = 1800 points.

Then they calculate the total maximum possible score. The maximum possible score for all three tables together is 1300 (Table 1) + 1500 (Table 2) + 1500 (Table 3) = 4300 points.

Then the percentage of the final score is calculated. The total score values from the previous steps are used and expressed as a percentage of the total maximum possible score. The total score is 88 (Table 1) + 70 (Table 2) + 80 (Table 3) = 238 points. The percentage score is calculated as follows: (Total score / Maximum possible score) * 100. Percentage score = (238 / 4300) * 100 = 5.53%.

Combining the values from the three proposed tables makes allows for creating a complex result that involves both the relevance of the monument virtual reconstruction and the existing mutual context of the monuments and the landscape. This result can be expressed, for example, using the total score or the scoring that combines values of individual criteria from the three tables. When creating this joint result, it is important to consider the importance of the individual criteria and their mutual influence on the overall context. For example, if the landscape context is a key factor for the evaluation, the importance assigned to criteria from that area may be higher compared to others. The result should provide a comprehensive view of the significance and value of the monument and its relationship to the surrounding landscape.

The mathematical process of obtaining a common result from the values of the three tables can be achieved in various other ways. One way is to weight and aggregate values from individual tables into a single overall score. Options for additional quantification evaluation are:

1. **Weighting of criteria:** Each criterion from each table can be assigned a weight based on its importance to the overall scoring. Weights can be determined subjectively, based on expertise or through statistical methods, such as Analytic Hierarchy Process (AHP).
2. **Aggregating the values:** Values of each criterion from each table can be aggregated into a single overall score by using weighted addition. Each value is multiplied by its respective weight, and then all the weighted values are added together to give the total score for each item.
3. **Normalizing the values:** Before multiplying the values by the weights, the values from each table can be normalized to the same range to prevent different values ranges from affecting the final score.
4. **Overall result:** After aggregating the weighted values from each table, the total score for each item is obtained. These total scores can then be compared and analyzed to determine the overall significance and relevance of the monuments virtual reconstruction and their mutual context with the landscape.

This evaluation can serve as a basis for decision-making in various areas of cultural heritage management and protection. In their planning and development, the classification can help identify areas of high cultural importance that need to be protected and preserved as part of regional development and urbanization. It can serve as part of the evaluation process in determining the protection measures and conservation practices, which can contribute to the sustainable conservation of cultural heritage. In the field of tourism, the classification can identify monuments with high tourism potential and help promote them as significant cultural destinations. Educational programs and popularizing activities can use the classification to raise the cultural heritage awareness and its importance to the society. In addition, the classification can also serve as a tool to support the financing of the monuments protection and conservation. The evaluation of cultural heritage according to the established criteria can serve as a basis for allocating funds for the monument protection, conservation and management.

The evaluations and calculations introduced above will be further demonstrated on specific case studies representing the practical application of this methodology. The goal is to verify the exactness and effectiveness of the evaluation and at the same time to provide a clear and systematic approach that can be used in other projects aimed at reconstruction and protection of monuments and historical landscapes.

Chapter 10

Case studies to uncertainty evaluation

To illustrate the above mentioned theory, three different cases were chosen from the practice to demonstrate the uncertainty relevance evaluation applied in digital reconstructions of cultural monuments in a historical landscape.

10.1 Water mill in the cadastre of Čebovce village

Introduction to the topic

A mill can be simply defined as a mechanical device for grinding grain, the result of which is flour of varying coarseness and quality. In Europe, the spread of mills is a significant indicator of civilization progress. Archaeological finds document their existence in the late La Tène period. Their improvement and expansion was related to the influences of Celtic and Roman culture. It is noteworthy that the use of millstones survived in the folk environment until the end of the 19th century, and in the Carpathian mountainous regions even until the middle of the 20th century. They spread in Europe between the 11th and 14th centuries. Description and drawing of a millstone is recorded in the work "Ten Books on Architecture" written by the Roman author Marcus Vitruvius. A simpler type of mill, the Greek mill, occurred in the area of Greece, today's Romania and Bulgaria, as well as in the northern part of Europe. Here, we cannot forget windmills, known as Dutch or German mills, named in relation to their most frequent occurrence (Tomovičová/Reichardtová 2023, 199; Botík/Slavkovský 1995, 365–366). In the course of time, milling technologies have evolved from simple water wheels to using water turbines, which increased their capacity and efficiency. The transition from folk to industrial mills is especially noticeable from the middle of the 19th century, when new technologies and materials began to be used.

Automation and mechanization of milling equipment significantly influenced the milling industry and contributed to modernization of production processes (Bartošíková 2023, 103).

In 1872, a total of 4,748 mills were recorded on the territory of present-day Slovakia, of which 4,527 were water mills (119). By 1894, the number of mills had decreased to 3,796, while 208 were registered in the Hont region. Suitable natural conditions and sufficient watercourses in Hungary and the Hont region supported their expansion and development for centuries. The first written records of water mills in Hont come from the 13th century, for example in Šahy (1235), Pláštovce (1282), Krupina (1316) and Vinica (1423). In the land register of Levice castle manor from 1554 the mills in Jabloňovce and Devičany are listed, where there were three mills, and two mills in Dekýš. According to the census from 1715–1720, the total number of mills in the Hont region was 123.

Under feudalism until 1871, only landowners (nobility, church) had the right to own mills, which right could be yielded to the city, village or any subject to rent. Mills were rented for three years to subject millers, who paid in kind, did carpentry work, or paid money. The millers were mostly craftsmen and their customers were mainly peasants from the surrounding villages. In small rental mills, millers mostly worked only with family members, while the working season depended on the amount of water in summer and frozen water in the winter.

The estates of the old Hont mills were stably located on the banks of streams and smaller rivers of Sikenica, Štiavnica and Krupinica (Urban 2010, 292).

At terminating 19th century, traditional mill wheels and stones began to be gradually replaced by more powerful motor drives and more modern cast iron grinding machines known as "roller mills". By 1906, eleven motor mills were already operating in the Hont region. The new, larger mills were used for industrial production of flour and other milling products, which they supplied to shops in cities and directly to the villagers. The old mills only ground imported grain into basic types of flour, peeled grain or groats. Owners and tenants of these small mills often could not afford to modernize their equipment, increase the production capacity or expand the product range. These factors led to the situation that, with the incipient 20th century, many mills in Hont gradually ceased to function and became useless. Since 1936, a nationwide law came to force to regulate their operation (Urban 2010, 293).

Mills in Slovakia are classified as technical monuments, and their protection has become a relevant part of the monument care since the inter-war period. Technical monuments, including mills, are characterized by their technical value and purposeful function, which are crucial for their existence. In addition to technical parameters, they have historical-documentary and urbanistic value as well. Protection and conservation of mills as technical monuments began to develop systematically in the 1970s, when a register of technical monuments and programs for their preservation began to be created (Tomovičová/Reichardtová 2023, 119).

To demonstrate 3D digital reconstruction, a defunct mill in the village of Čebovce was chosen, as it seems very interesting. In the past, the mill was located in the historic region of Hont (the present-day Veľký Krtíš district). The village centre is located at an altitude of 225 m, while altitudes within the village outer limits range from 200 to 567 m. The village is situated at the southeastern foot of the Krupinská planina plain, where the valley of Čebovský potok stream passes into the basin of Ipeľská kotlina. The southeastern part of the village district is situated in the basin and has hilly character with Neogene deposits, while its northwestern part on the Krupinská planina plain is more rugged and has upland character built of andesite tuffs. There are brown coal and basalt deposits in the village district. The district is involved in the climatic region of continental basin climate (Michalová/Michal 1980, 260). The soil cover consists of brown soils, illimerized soils and brown forest soils. Agricultural land occupies a total area of 792 ha, of which 550 ha is arable land, 54 ha is meadow and 112 ha is pasture. The forest land covers 728 ha. Forests (oak and beech) are found mainly on the slopes of Krupinská planina plain.

In the village, Slavic settlement finds dated to the 8th century were found. The first written record of the village dates back to 1330. The village had the character of a subject peasant settlement, which belonged to the Zichy family and the Balasso family during the Middle Ages. In the 20th century the property was owned by the Somogyi family. The inhabitants were also engaged in viticulture. In 1715, the village had a mill and 68 households, in 1828 there were 112 houses and 674 inhabitants (Michalová/Michal 1980, 261).

The digital visualization of the mill in Čebovce village was created to demonstrate the reconstruction of a completely extinct object, on which the proposed classification system of uncertainty has been applied to determine display relevance.

Source base of digital reconstruction

For the selected object of defunct mill in Čebovce village, a digitized document was used at first. The plan called Csabensium Molarum is a valuable historical document capturing the technical and construction solutions of the water mill in Čebovce from 1790. This plan, created by Johannes Scmeidler-Cameral, belongs to the file Div. VIII. under number 46 and is deposited in the National Archives in Hungary (maps.hungaricana.hu; S_12_-_Div._VIII._-_No._46.). The plan is 51 x 69 cm in size and is handwritten, partially coloured, which makes its technical elements clear and comprehensible. Its scale is 91 mm, which provides a detailed and accurate view of the building and its components. The plan is drawn on paper and is partially coloured. A brief descriptive text states that it is an original manuscript that contains cross-sections of the dam (Fig. 31). The plan includes several important parts that show individual aspects of the water mill in detail (Fig. 32).

The floor plan shows the overall layout of the mill, including interior and exterior spaces, and clearly shows how the individual parts of the mill are arranged. The cross-sections provide a detailed view of the dam and mill canals, revealing the technical solutions and the dam construction. These sections are crucial for understanding the mechanics and functioning of the mill and, of course, for its reconstruction. The plan also shows the mill canal (malomcsatórnal) and ditch (arak), which are integral part of the mill water system. Details of the bridge (híd) and plug (zsilip) are involved in the plan, as it is important for the water flow regulation and management.

In the case of reconstruction of defunct architectural-historical monuments, analogies from other similar conserved buildings of the same type play a very important role. Comparison with other preserved or documented mills from the same period and region can provide valuable information about the architectural and technological solutions used at the time. This helps fill in missing parts of the plan or clarify unclear details. In this case, published sources for these technical monuments were used (Bartošíková 2023 et. al) along with historical photographs showing similar buildings. If some details are known about other types of mills (e.g. windmill, manual mill), this information can be used to understand the universal principles of mill functioning. For example, the mechanisms for transferring power from water to millstones may be similar in different types of mills.

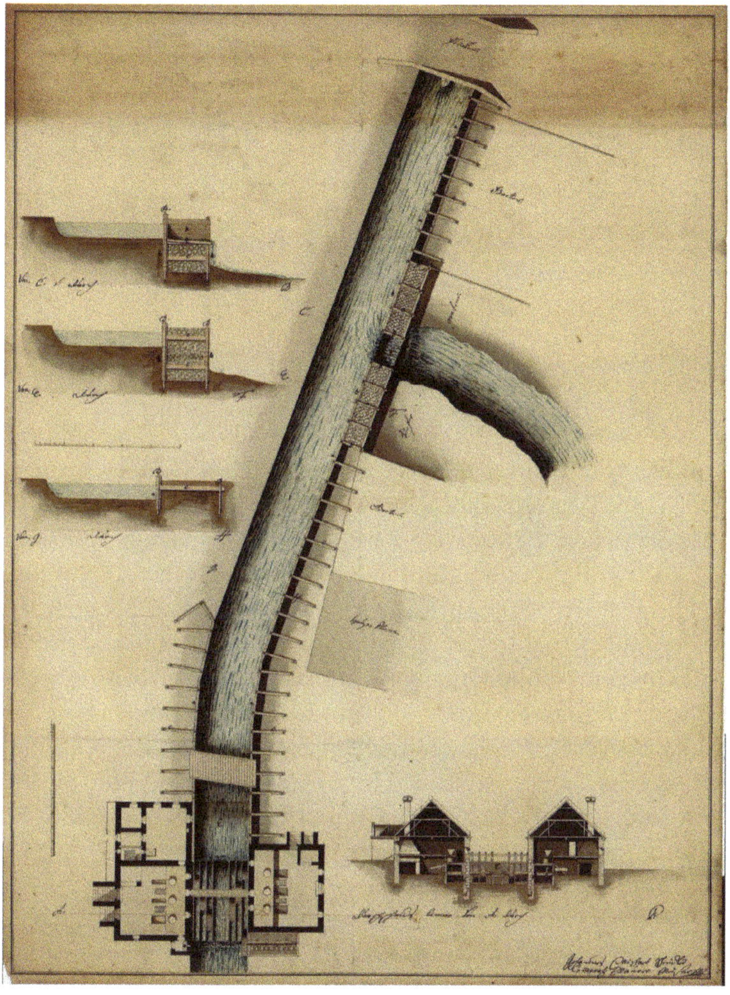

FIGURE 31

Its location in the current landscape environment is also relevant. Since it is a defunct building, it is advisable to verify its original location in the field by means of field research. First of all, historical maps from the late 18th and 19th centuries need to be georeferenced with present-day maps, including

FIGURE 32

materials obtained by the LiDAR method. Based on this method, two potential places where the reconstructed mill could be situated were identified in the field. The first one is located 1.5 km northwest of the urban space of original village, and the second is located within the village limits near the original road to Kosihovce village. In both cases, the presence of this structure is difficult to verify, as the surface survey has its limited possibilities. The defunct building structures are barely visible on the ground and their layout is very fragmentary. Inundation and erosion processes lasting more than a century in the Višňový potok floodplain have to be considered as well, as they could have changed the original floodplain appearance with the mills situated there beyond recognition. The mill identification could be solved in future by several destructive archaeological excavations along with further analyzes of historical sources and maps.

Reconstruction methods and procedure

Reconstruction of the mill in Čebovce was created in Blender program. Its primary source was the above-mentioned Csabensium Molarum plan from 1790. The modeling process began with thorough preparation and analysis of the documents. Before modeling itself, all parts of the plan had to be examined in detail. Then the entire plan was digitally downloaded in high resolution in order to work with all the details in individual parts of the plan. It also contains cross-sections of buildings and the water flow. These sections were taken out and vertically inserted directly into the plan locations. After importing the reference images into Blender, setting the scale correctly was crucial. The plan uses a scale of 91 mm. The units in Blender were set to meters and dimensions of all objects were recalculated according to this scale. This step involved calibrating the reference images so that all dimensions in Blender 3D space corresponded to actual dimensions of the reconstructed mill (Fig. 33).

Subsequently, the reference images were imported into Blender and placed in an orthographic view (top, front, side), which allowed for exact modeling of individual mill parts according to the plan. The images have to be properly aligned and scaled to match each other in scale and position.

At the start, basic ground plan of the mill was modeled. Using Blender tools, 2D meshes were created that copy the mill outline according to the ground plan. The "Extrude" tool was used to create the base walls and floors, resulting in a three-dimensional foundation for the building. This step involves creating of the main building and adjacent structures, while caring to be accurate according to the original plans (Fig. 34).

After creating the basic ground plan, walls and interior spaces were modeled. The "Loop Cut" and "Knife" tools were used to create necessary partitions and internal walls. In this step, exact locating of all rooms and internal partitions according to the ground plan is important. Next, attention was

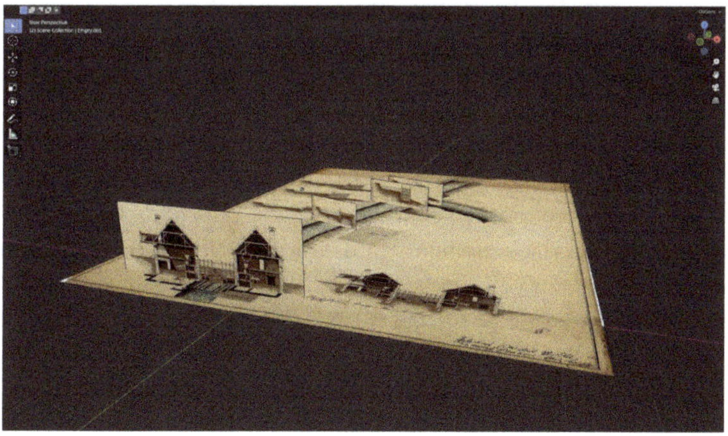

FIGURE 33

FIGURE 34

focused on exterior details. Windows, doors and roof elements were modeled using reference images to ensure these components were proportionally correct and historically accurate. Tools like "Inset" and "Bevel" helped create realistic details that add authenticity to the model.

After the basic structures and exterior details had been completed, we moved on to modeling the mill mechanisms, such as water wheels, shafts and stoppers (Fig. 35). These elements are essential to understanding how the mill works, so we used tools like "Spin" and "Array" to create repeating elements, e.g. waterwheel blades. We made sure that these mechanisms are accurate and functional according to historical records.

The next step is modeling of the cross-sections and technical details that are shown in the plan. These include the mill dam, bridge and ditch details. "Boolean" tools were used to create precise cut outs and openings, which allows us to model these elements in detail. Texturing of these details was created by using historically accurate materials to achieve an authentic look. In this case, materials and textures had to be obtained secondarily from artefacts meeting the requirements for structure and colour. The materials, such as old wood, wet wood, stone or plasters were obtained in the field or on available internet portals. Secondary raster maps (normal maps, depth maps, spectral maps, etc.) were also created for all textures for more realistic

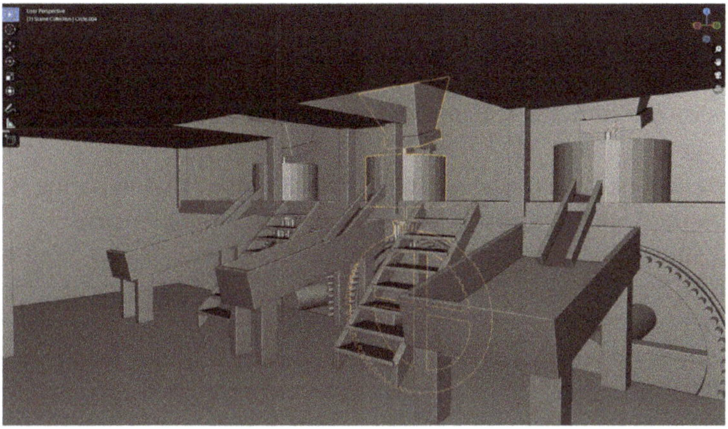

FIGURE 35

illusions of the final material. After completing the modeling of all structures and mechanisms, textures and materials were applied to the model. The textures were applied by using UV mapping to achieve the most realistic look possible and proper covering of the model surfaces.

In the case of this reconstruction, no 3D terrain was created, as it was not possible to identify the place where this mill originally stood accurately.

Finally, the scene lighting was set up. Natural daylight was used to simulate sunlight and several light sources to highlight interior details. Exterior light with sky was created by using HDR texture. Additional vegetation was created in the scene in the mill vicinity, which, however, is just an aesthetic element. It is not based on any relevant and verifiable sources. Rendering the scene in Blender using the Cycles render engine provided high quality images of the reconstructed mill.

Evaluation of uncertainty visualization

Evaluation of uncertainty visualization was preceded by filling the evaluation table in the process of analysis and synthesis of available and already used sources. Therefore, the relevance of individual parts of the building could be evaluated and visualized within the modeling process.

Criteria in the evaluation table were filled as follows:

Preserved: no part of the mill building has been preserved to this day in visible above-ground structures or in fragments. Therefore, it was not possible to start from any authentic construction elements.

Architectural-historical or archaeological research: no research carried out

Detailed location of the object: not feasible due to the building demise

BIM implementation: not feasible due to the building demise

3D digital documentation: not feasible due to the building demise

Historical technical plans and failures: at the time of reconstruction, only one building plan from 1790 was available

Vedutes or historical drawings: no one available at the time of reconstruction

Historical photographs: no one available at the time of reconstruction

Direct analogy: no direct analogy available, but analogous buildings of the kind were used and are still preserved or documented in almost the entire building mass.

Written sources regarding the object shape: no one available at the time of reconstruction.

Texture: since the building was not preserved in physical matter, textures needed to be obtained through photographic documentation of textures on preserved buildings or building elements. Textures that could not be documented were replaced with ready-made ones on available websites.

Material: The procedure of obtaining materials was the same as of textures, with the difference that materials had to be adapted to the lighting conditions and characteristics of individual models.

Spatial context in situ: it was not possible to verify the mill building exact layout in the field according to the historical Csabensium Molarum plan, as the identified surface structures were barely visible. Therefore, we assessed that the spatial context cannot be precisely identified for now.

Evaluation of the data entered in Table 6. was as follows. The final score from the table after adding the individual attributes is 200 points out of total possible value of 1300 points. In percentages terms, the certainty degree of the digital architectural-historical visualization of the defunct mill in Čebovce is 15.4%.

Table 6 Evaluation of uncertainty visualization in the reconstruction of defunct mill in Čebovce village. Created by D. Bešina.

	100	50	0
Preserved			
Architectural-historical or archaeological research			
Detailed object location			
BIM implementation			
3D digital documentation			
Historical technical plans and failures			
Vedutes or historical drawings			
Historical photographs			
Direct analogy			
Written sources to the monument shape			
Texture			
Material			
Spatial context in situ			
Total			

Final reconstruction

At first glance, final digital reconstruction of the mill in Čebovce gives a very faithful and realistic image of the building in its original environment (Fig. 36). However, the certainty degree is very low in this case, which de facto may reduce the credibility and authenticity of the visual representation. Therefore, in this case, setting off the building parts with different levels of preservation and credibility on the digital model is definitely pointless, as none of it exists anymore. Nevertheless, this reconstruction can be used as a supplement to possible future presentation outputs. It can also be used in the case of future archaeological research on the site, so that it would be possible to add individual investigated parts of the building to a virtual model and thus make it more accurate, which would increase the uncertainty degree values in the table and its overall evaluation exponentially. It follows from the above that this evaluation is not unchangeable, but it could be dynamically changing according to the new knowledge or sources added.

The reconstruction also confirms another important fact for the general practice. A higher level of realistic visualization of computer reconstructions does not automatically mean that the display level is significantly authentic

FIGURE 36

and accurate. The possibilities of rendering digital materials, textures, models and atmospheric phenomena are currently at such a level that even completely extinct buildings or landscape units can be displayed at a very realistic level. However, the uncertainty visualization is a matter of discussion, which is of fundamental importance for scientific purposes.

10.2 Manor house and historical park in Dolná Strehová village

Introduction to the topic

The village of Dolná Strehová is located in the northeastern part of the Ipeľská kotlina hills, in the river valley of Tisovník. The area, known as Strehovská pahorkatina, is characterised by hilly relief and is formed by Neogene sediments and partly by tuffs. Its steeper slopes, which are not used for agriculture, are mainly covered with oak and acacia forest vegetation. The territory is rich in thermal mineral springs. The village centre is located at an altitude of 185 m; the altitudes within the village limits vary from 170 to 311 m. Building stone - andesite tuffs - was quarried there in the past. Regarding its climate, the area can be included in the climatic region of the continental mountain slopes climate, in the climatic region of Krupinská planina plain. The soil cover consists mainly of brown forest soils, brown soils, floodplains and alluvial soils. Agricultural land (1392 ha) occupies the majority of the village district. Of this area, arable land is 864 ha, meadows 171 ha and pastures 324 ha. Forest land covers an area of 604 ha. The forest vegetation of acacia and oak trees covers mainly steeper slopes (Michalová/Michal 1980).

The oldest settlement of the village has been proven by archaeological finds to the Bronze Age, the La Tène period and the Early and Late Middle Ages (Beljak/Mitáš 2013; Furmánek 2000). In Slovak language, the name of the village is probably etymologically related to its original function of guarding, which was its assignment probably performed in the 10th and 11th centuries. Significance of guarding is also reminded in the name of Strážna hora (Guarding hill), which is located on the southeastern edge of the village limits, in the close vicinity of river Ipeľ. Guarding measures in this area were always important here, as the Ipeľ river valley suddenly narrows here and forms a corridor, which in the past was an important way connecting the Ipeľ river valley with the Slaná river valley and heading to Miskolc. The first written record dates back to 1245, where a certain Aba and his relative Hont, son of Lazar, and Dedáč, son of Voč are mentioned (Alberty 1989, 133–134). In the 14th century, a church is also documented here. In 1423, Stregowa is mentioned among the toll points of the Novohrad County. In the 15th century, František Madáč became the owner of Strehová, but some disputes among several noble families concerning the property ownership are recorded in written sources of this period. In the Novohrad County portal inventory from 1549, Štefan and Martin Madáč are listed as of the village owners (Adamove 2015, 88). The place was a serf village of several families: the Ztergua family, the Zechen family, and the Madáč family from 1430. In this period, the power of Countess Néhai Wenchteim, nee Forgách, was partly concentrated here. In 1552, the village was occupied and destroyed by the Turks; in this period it was a border settlement of Sečany and Fiľakovo sanjak (Alberty 1989, 155). At the beginning of 17th century, Dolná Strehová became the place of metropolitan congregations. In the first half of 17th century, a Latin high school was founded in the village, which was also supported by the Madáč family. In 1693, the well-known polymath Matej Bel of Očová studied there. The school was located near today's Evangelical church. Originally a late-Renaissance church from 1652, it was rebuilt in the Baroque-classicist style in 1819. The Church of the Virgin Mary´s Assumption, originally classicist from the second third of the 18th century, was adapted in the art nouveau style in 1903, which was especially reflected in its historicizing facade (Güntherová 1967, 316).

In 1809, Alexander Madáč bought a manor under construction, originally built in the Baroque style, from the descendants of Peter Révai. He subsequently rebuilt this mansion according to the new classicist style, creating a magnificent residence. The Révai family decided to build a new mansion in Dolná Strehová, because the previous one was already old and did not meet their needs. Although it was younger than the Madáč manor house, it was probably smaller and the family status required a more representative residence.

The original manor house of the Révai family, which they decided to demolish, was situated somewhere near the current residence and the Evangelical church, maybe in the same place. This assumption is based on a canonical visitation in 1624, in which is stated that Ján Rimaj attached a part of the church to his manor house. After Rimaj's death, the manor was inherited by his daughter Mária, who brought it into marriage with the Dulaj family. Later, the residence passed into the hands of the Benická family, and finally it became the property of Peter Révai, who married Zuzana Benická. The Révai family inherited the manor until the last owner, from whom it was bought by Alexander Madáč. The existence of the old house is confirmed not only by the canonical visitations from 1624, but also by Josephine mapping from 1763–1787. During the recent reconstruction of the manor house, a remnant of the bastion that was part of the original building was even found (Adamove 2015, 92–93). The predecessor of today's manor house burned down in 1717 and 1758. The old manor house served as a granary until it was demolished in the 1930s by Dr. Lazár, the husband of Flora Madáč, who was the last Madáč family member. At the end of the 18th century, Alexander Madáč started building a new family residence, which was completed by his son Imrich sr. in 1823. The current manor building with corner towers was built in 1853. The three-winged U-shaped layout with a central buttress and an inclined tower refers to older architectural traditions of fortified mansions. A typical manifestation of early classicism is the stucco ornamentation and the overall arrangement of the facades (Güntherová 1967, 315).

Thus, the manor house built by the Madáč family in the Rococo-classicist style is one of the important monuments in the village of Dolná Strehová (Fig. 37 and 38). The mansion is closely connected with the well-known writer Imre Madách, the author of "Tragedy of Man". Most of his significant literary works were written here. From the point of view of cultural landscape research, the English park near the manor house is interesting as well, which was described by Mocsáry in his monograph on Novohrad County from 1826. Alexander Madáč had a large English park built around the manor house with trees imported from many distant countries. The manor owners later built a tennis court and a bowling alley in the park. At the end of the park, there is still the tomb of Imre Madách and his family with a statue of Adam. Part of the park has been preserved in the Rákyti area, where old trees are still standing, reaching a height of 18 to 35 meters and 400 years old (Fig. 39). The park underwent relevant changes in the second half of the 20th century, when a school and a gymnasium were built on its premises, which significantly changed the original appearance and elements of this territory (Adamove 2015, 92).

FIGURE 37

FIGURE 38

FIGURE 39

Source base of digital reconstruction

The aim of virtual reconstruction is to show the Madáč family's manor house and historical park as they looked in the second half of the 19th century. For this purpose, various types of sources were obtained, which had to be converted into digital form. Regarding the reconstruction, the fact that the building still stands in its complete structural form with minor structural modifications or changes is significant. After its extensive reconstruction, it currently functions as a museum. Positioning the manor house on its original place and in its original building material enabled selected digitization methods, such as single-frame photogrammetry and aerial photogrammetry with UAV equipment, to be applied. Perpendicular orthographic images of all frontal parts including the facade with the buttress in high resolution were obtained by means of single-frame photogrammetry, which later contributed to a better rendering of building details such as window sills, etc. The single-frame photogrammetry was taken from the ground with a digital camera on a monopod and was supplemented by images from UAV (DJI Mavic Air 2 drone), as the height of the building did not allow capturing all facade parts from the ground. Another digitization method of aerial photogrammetry was implemented with the drone, the aim of which was to capture the complete 3D geometry of the mansion from a height. The obtained data made it possible to create a spatial model in the form of a point cloud, which was modified into a polygonal model.

The manor house reconstruction required collecting of mainly visual sources such as photographs and historical drawings or paintings, substantial part of which is well published in newer publications (Adamove 2015) and older sources, especially from the late 19th and early 20th centuries. Most of them were photographs on period postcards, hand drawings or paintings (Borovszky 1911). Valuable images can also be found in Imre Madách museum exposition right in the mansion. All these iconographic sources capture the mansion with the adjacent park at the time it is intended to be reconstructed, which makes them a relevant source for virtual reconstruction.

Another relevant source is historical maps, especially regarding the house location in the original historical park. The first three military mappings of Old Hungary made within the time period of ending 18th century to the second half of 19th century (mapire.eu) were important as well. We used also maps from the second half of 20th century for our purpose, especially historical aerial photographs from the 1950s (Historical orthophotomap TUZVO) and basic maps. In order to create spatial model of the landscape around the manor house on which the original vegetation cover of the historical park was reconstructed, LIDAR data needed to be downloaded from the ÚGKK SR portal in the form of raster DMR image in .tiff format. Unfortunately, map or image sources do not capture the historical park to a sufficient extent, which is significantly reflected in the uncertainty level evaluation. We used the only reliable image source, which is a hand-made layout of the entire park, the author of which is Imre Madách himself. The degree of detail of the park sections is not high, nor is the scale, but it is drawn by a person who lived there and could draw the disposition and distribution of the park elements as he saw it with his own eyes (Fig. 40).

All map data were georeferenced in the QGIS program in order to obtain more accurate spatial situations of the mansion and parts of the historical park. In this program our own maps were also drawn, which made it possible to create a road network or vegetation cover during modeling. All created maps were applied to the spatial model of the defined landscape area. The landscape model was exported in .stl format based on the DMR raster base in the QGIS program (Fig. 41).

Reconstruction methods and procedure

Modeling of the manor house in Blender starts from source processing, which involves orthographic imagery and point clouds obtained from photogrammetric procedure. This data is imported into Blender as the overall model basis. Orthographic images provide accurate perpendicular shots of all the building frontal parts, while the point cloud from aerial photogrammetry offers its complete 3D geometry. The point cloud was processed in Metashape and exported to .ply format. Orthographic images from single-frame photogrammetry were processed in SimPhoto.

The source import process involves launching Blender and creating a new project. The "Import Image as Planes" function is used to import orthographic images as reference planes, which are then placed in Blender space to match the real proportions of the mansion. In this case, we also make sure that the mobile images and the future model are on 1:1 scale. The point cloud is imported using the "Point Cloud Visualizer" plug-in and set as a reference for accurate building geometry, thus creating the building outline based on the point cloud to ensure correct orientation and scale.

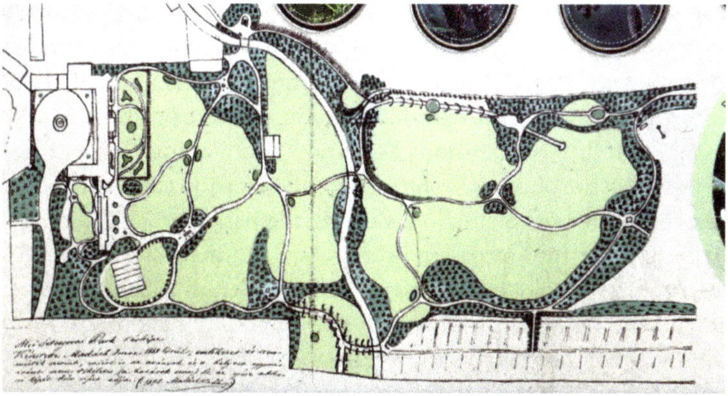

FIGURE 40

FIGURE 41

After importing the digitized image sources, the mansion basic geometry is created. Historical depictions of the manor house that serve as templates necessary for completing the building parts were opened on the program side screen. The process starts from creating the building basic shapes using primitive objects such as cubes and cylinders. These objects are positioned and scaled according to the orthographic images and the point cloud to create the main structure of the building including walls, roofs and main facades. The mansion overall modeled geometry is refined by editing auxiliary photos in the fSpy program. This program creates an accurate perspective of digital photography in three dimensions X, Y and Z from a single view (Fig. 42.).

After editing, the photo is imported into Blender and can then be modeled without compositional distortion. Editing the basic model shapes includes using the Extrude tool to create protrusions and depressions in walls and facades, the Bevel tool to round edges and add details such as cornices and facing, and the Loop Cut tool to add new cuts to the geometry for better control over the model shape and details.

After creating the basic geometry, detailed modeling follows. This step includes modeling all facade details according to orthographic images and templates. Details such as windows, cornices, doors and buttress are modeled using modeling tools, paying attention to the original construction accuracy and veracity (Fig. 43.). Detailed modeling also involves creating of textures, which are obtained from photographs captured by photogrammetry and edited in a graphics program (GIMP). The textures are applied to the model using the "UV Unwrapping" tool, which allows the textures to be placed precisely on individual model surfaces.

FIGURE 42

FIGURE 43

The model final editing includes checking based on reference images and point clouds, correcting proportions and details, adding small details such as gutters, chimneys and other architectural elements that increase the model feasibility.

After importing the landscape model into Blender, the process of detailed modeling of the historical park begins. The already made mansion is placed on this terrain in the required scale. The basic imported terrain is modified as needed using the "Sculpting" tool, which enables the landscape shaping according to altitude data from LiDAR. The terrain is adjusted to correspond to the historical maps and the situation plan. Many of the landscape structures captured by LiDAR are of younger data and come from the recent decades. Then, the road network is modeled according to the referenced map data from QGIS. Using the "Grease Pencil" tool, roads and paths are drawn in Blender, which are then transformed into 3D objects.

The park vegetation is modeled using the "Particle System", which allows for grass, trees and bushes to be distributed over the modeled terrain. Plant types are selected from the Botaniq plug-in library installed in Blender. Models of trees and shrubs are created and then randomly distributed around the park. Unfortunately, exact plant species cannot be imported into the scene, as today it is not possible to say how many and what trees or ornamental plants were planted in the park. Important trees and plants that are captured in historical images and modeled in detail deserve special attention. Based on the site plan, relevant components of the park, such as benches, fountains and statues, are placed along with the small park architecture captured in the photographs.

The completed modeling of vegetation and other elements is followed by applying of textures. Textures are obtained from photographs of historical materials and edited in the GIMP graphics program. Using the "UV Unwrapping" tool, the model surface is decomposed into 2D space, which allows for the textures to be placed accurately on the model surfaces. Applying the textures by using "UV Mapping" ensures that the textures exactly match the shape and details of the model, thus achieving a realistic look (Fig. 44).

The final step is setting the lighting and rendering. Realistic scene lighting is set using HDRI maps and light sources. After the lighting is complete, final renders are created from different angles and perspectives for the model presentation.

Evaluation o uncertainty visualization

Similarly the first case reconstruction, the evaluation of uncertainty visualization was preceded by filling the evaluation table already in the process

FIGURE 44

of analysis and synthesis of available sources. Therefore, the relevance of individual parts of the manor house were evaluated and visualized as part of the modeling process. Number of available sources directly affected the data entry and the resulting table score.

Criteria in the evaluation table for architectural-historical objects were filled as follows:

Preserved: the manor house has been preserved and reconstructed in all its mass. The only changes are visible on some architectural elements. The interior is almost in its original layout, modified for the needs of museum presentation.
Architectural-historical or archaeological research: minimum
Detailed location of the object: at no disposal in the time of reconstruction
BIM implementation: not feasible due to insufficient technical possibilities and its complete absence in monument care practice in Slovakia
3D digital documentation: only partial in the exterior using the method of aerial photogrammetry with UAV equipment
Historical technical plans and failures: no one available at the time of reconstruction
Vedutes or historical drawings: several period drawings of the manor house itself from the second half of the 19th century were available
Historical photographs: the number of photographs from the end of the 19th century to the present was sufficient
Direct analogy: in our search for direct analogies, we focused on contemporary mansions from the region and the wider surroundings. In this case, however, these analogies were only supplementary, as the building is relatively well preserved.
Written sources regarding the object shape: no one available at the time of reconstruction.
Texture: obtained from the object directly and modified accordingly
Material: obtained from the object directly and modified accordingly

Evaluation of the data entered in Table 7. was as follows. The final score from the table after adding the individual attributes is 750 points out of total possible value of 1300 points. In percentages terms, the certainty

Table 7 Evaluation of uncertainty visualization in architectural-historical visualization of the manor house in Dolná Strehová village. Created by D. Bešina.

	100	50	0
Preserved	■		
Architectural-historical or archaeological research		■	
Detailed location of the object			■
BIM implementation			■
3D digital documentation		■	
Historical technical plans and failures			■
Vedutes or historical drawings	■		
Historical photographs	■		
Direct analogy		■	
Written sources about the object shape			■
Texture	■		
Material	■		
Spatial context in situ	■		
Total		**750**	

degree of the digital architectural-historical visualization of the manor house in Dolná Strehová is 57.7 %.

The evaluation of 3D reconstruction of selected historical landscape in 3D virtual space was also applied to the historical park near the manor house in Dolná Strehová. The evaluation attributes were gradually inserted into the proposed table.

Evaluation criteria were added according to the current state of research and possibilities of the source base:

Changes in geo-relief: no remarkable changes in the terrain geo-relief
Changes in micro-relief: changes in the terrain micro-relief are present
Changes in the original landscape cover: remarkable changes in the selected landscape cover in the park are present
Research on anthropogenic phenomena on the territory: no research known at the time of reconstruction
Research on environment of the territory: no research known at the time of reconstruction

Anthropogenic structures: no remarkable anthropogenic structures in the park apart from several short road sections that, however, are not corresponding with original road system in the park

Historical mapping of the territory on a small scale: Three cadastral maps from the second half of the 19th century have been preserved, which do not capture the park interior, but only its territorial boundaries. The plan of the park drawn by Imre Madách is still preserved.

Historical mapping of the territory on a medium scale: three Old-Hungarian military maps have been preserved, where the park boundaries are visible.

Historical mapping of the territory on a large scale: large-scale maps are irrelevant for the reconstruction of the park, as it is a rather small area.

Historical aerial photographs: digitized aerial orthographic photographs from the 1950s preserved.

Detailed cartographic localization of the territory (LiDAR DTM DMR): data obtained by LIDAR are available in the point cloud or in raster DMR map.

3D digital documentation (drone): for comparison, the park was photographed with UAV device. However, its mapping in the form of perpendicular orthographic images or 3D recording was not done, and therefore we evaluate this attribute with only half the value.

LIM implementation: no analyzes known at the time of reconstruction.

Vedutes or historical drawings: during the reconstruction, only two drawings of the park adjacent to the mansion were used, which helped only partially.

Historical photographs: the park is captured on contemporary photographs only marginally.

Written sources: only a brief description of the park's equipment and some tree species are available.

Table 8 Evaluation of uncertainty visualization of the historical park adjacent to the manor house in Dolná Strehová village. Created by D. Bešina.

	100	50	0
Changes in geo-relief	■		
Changes in micro-relief		■	
Changes in original landscape cover			■
Research on anthropogenic phenomena			■
Research on environment			■
Anthropogenic structures			■
Historical mapping – small scale	■		
Historical mapping – medium scale	■		
Historical mapping – large scale			■
Historical aerial photographs	■		
Detailed cartographic localization (LiDAR, DTM, DMR)	■		
3D digital documentation (drone)		■	
LIM implementation			■
Vedutes or historical drawings		■	
Historical photographs		■	
Written sources		■	
Total		**750**	

Evaluation of the data entered in Table 8. was as follows. The final score from the table after adding the individual attributes is 750 points out of total possible value of 1600 points. In percentages terms, the certainty degree of the digital visualization of historical park in Dolná Strehová is 46.9 %.

At this point, another example of the evaluation of mutual landscape context of the manor house in Dolná Strehová has to be mentioned. It is necessary to point out that the landscape context of selected monument or monuments in the cultural landscape is, to say the least, complicated to assess. The evaluation should be based on deeper interdisciplinary research, as several scientific areas are reflected in it. Therefore, the lower rating is, in essence, only an example for future potential application in practice.

Table 9 Evaluation of mutual context of the manor house and adjacent historical park area in Dolná Strehová village. Created by D. Bešina.

Context	Criteria	100	50	0
Historical	Important events		■	
	Chronological context			■
	Historical relevancy		■	
Architectural	Unique style and techniques		■	
	Materials and technical details			■
	Proportions and aesthetic value	■		
	Connection with the residence urbanism	■		
Landscape	Integration with natural environment			■
	Important site in the landscape	■		
	Access and sightseeing		■	
	Direct link with other anthropogenic structures in the vicinity	■		
	Direct link with other architectonic objects in the landscape		■	
Social and cultural	Residents participation		■	
	Importance for local community	■		
	Cultural and religious associations	■		
Economic	Contribution to local economy	■		
	Potential for tourism	■		
	Investments in protection and restoration		■	
Total			**1050**	

Evaluation in Table 9. is only a model within the presented study, but it can nevertheless be used as a working evaluation of the landscape context of monuments in landscape environment. Total rating of the attributes entered in the table is 1050 points out of a maximum of 1800 points. In percentages terms it means that in the selected territory of historical park with anthropogenic objects in Dolná Strehová, the context degree of monument and historical landscape is preserved at 58.3 %.

All the evaluations obtained can be comprehensively summarized in the following Table 10.

Table 10 Summarizing evaluation of uncertainty visualization degree of virtual reconstruction of manor house and adjacent historical park in Dolná Strehová. Created by D. Bešina.

	Final evaluation in %
Evaluation of certainty degree of manor house digital reconstruction in Dolná Strehová	**57.7 %**
Evaluation of certainty degree of historical landscape digital reconstruction	**49.9 %**
Evaluation of preserved mutual landscape context of monuments and landscape	**58,3 %**

FIGURE 45

Final reconstruction

The summarizing table enters relative data that can be used in future research or in solving complex issues associated with preserving the historical values of a monument and the territory in which it is located. The evaluation obtained can be described as relatively good. This means that the rate of virtual reconstructions is of high quality and informative value (Fig. 45. – 47.). Quantification of the certainty degree of the manor house and park reconstruction will contribute to a better definition of what can be considered a monument authentic essence and what has disappeared in the course of historical events. In addition to professional scientific outputs, these data can also be used in presentation activities, as even the lay public should in real recognize the reconstruction certainty degree in order to realize how much “has remained” of the monument's values.

The reconstruction provides the viewer with several realistic glimpses of the park and the mansion. The outputs are focused only on those park areas that could best be reconstructed according to the available sources. The visualization credibility is enhanced by realistic vegetation and lighting of the scene. The outputs can be presented in professional articles and publications or directly in the premises of manor house museum. In this case, the possibility of constant improvement of the virtual scene, which is directly proportional to the amount of new knowledge obtained through cabinet or field research, is an advantage.

Final results are presented in Figure 47, highlighting key aspects of the visualization

FIGURE 46

FIGURE 47

10.3 Reconstruction of the vineyard landscape in the cadastre of Nenince

Introduction to the topic

In Slovakia, viticulture has a long and rich tradition, which is reflected in many aspects of local culture, especially in regions predisposed to this form of farming. Wine-growing areas in the Slovak wine-growing region are classified in zone B, similar to areas in Austria and Moravia in the Czech Republic. The exception is the Tokaj wine-growing region, which together with Hungarian Tokaj regions belongs to zone C I (European Commission 2019; Némethová/Krajči 2021; Regulation of the European Parliament and Council No. 1308/2013). Viticulture in Slovakia has been developing at least since the Top Middle Ages, which led to the formation of winegrowing regions. This process continued until the 18th century, when the wine producing areas we know today were created (Kazimír 1986). The countryside character was differentiated, which stimulated emerging specific wine-growing landscape, which underwent several changes due to climatic and geological conditions, to which the grape varieties reacted differently.

The historical vineyard landscape was characterized by small-block vineyards, which were often combined with orchards, arable land or pastures (Drábiková 1989; Slobodová Nováková 2020). Planting of fruit trees in vineyards was common, which complicates the precise defining of their extent. When determining these territories, the temporal aspect of their planting has to be taken into account, for example before or after the collectivization (Špulerová/Štefunková/Dobrovodská 2017).

Mechanization in the 20th century moved vineyards to flat terrain. At the turn of the 19th and 20th centuries, a viticulture calamity caused by an epidemic of phylloxera (*Viteus vitifoliae*) broke out, which led to decrease in planting of new vine roots, introducing new varieties and moving the vineyards to sandy plains (Jurkovič 1969).

In the second half of the 20th century, there were fundamental changes made in the purposeful land use, modernization of plantings and modification of sloping terrain. The goal was to refine the vine-growing areas and create more efficient wine-producing areas, which led to liquidation of unpromising vineyards (Drábiková 1989; Slobodová Nováková 2020). As the vineyards and orchards were deserted, extensively used landscape of overgrown orchards and woodlands gradually appeared (Chrastina 2023).

Currently, the demand for wine products is growing, which is affecting the overall character and preservation of the traditional vineyard landscape. Thanks to wine, wine tourism is developing, which is significant economic component for rural businesses (Věžník/Kyselý 2019). Therefore, preservation of original objects and structures of viticulture landscape is a topical issue.

In Slovakia, there are legislative instruments enabling protection of valuable wine-growing areas. However, this process is not easy and requires cooperation and consensus of several legal documents, authorities and organizations. Adoption of the European Landscape Convention (Florence 2000), signed by the Slovak Republic in 2005, increased the emphasis on research, documentation and protection of cultural landscape in Slovakia (Dvořáková 2012). The Convention integrates the protection of nature and cultural monuments (Štefunková/Krnáčová 2020). Traditional agricultural landscape, including historic vineyard areas, represents a synthesis of nature, history and the settlement culture, protection of which is partially ensured by the Ministry of Culture through Act No. 49/2002 Coll. on the Protection of the Monument Fund and by the nature and landscape protection authorities according to Act No. 543/2002 Coll. on the Nature and Landscape Protection (Špulerová/ Štefunková/Dobrovodská et al. 2017).

Natural anthropogenic transformation processes caused the disappearance of many historical landscape components, hence documenting of remains of these objects through interdisciplinary cooperation directly in the field is necessary.

This exemplary study is aimed in reconstructing a defunct historical vineyard landscape in the selected territory of the village Nenince cadastre in the middle of 19th century and in evaluating the certainty degree of the historical landscape reconstruction.

The village of Nenince is located in the geomorphological unit called Čebovská pahorkatina hills and with the area of almost 17,383.5 ha is one of three geomorphological parts of the Ipeľská kotlina hills, which belongs to the sub-unit of South Slovak basin. This area is located north of the Ipeľ River in the district of Veľký Krtíš. In the east, it is connected to the Pôtorská planina plain and the Hontianske Terraces, which line the undulating hilly landscape from the south. From the north, this area is joined by the Slope of Modrý Kameň, belonging to the sub-unit of Krupinská planina plain (Michalová/Michal 1980; Vass/Konečný/Šefara 1979; Kočický/Ivanič 2011). The district relief is hilly with flat ridges, situated on Neogene deposits. The village area belongs to the climatic region of continental basin climate, to the climatic region of the Ipeľ river basin. Its soil cover consists of brown soils, meadows and alluvial soils. Agricultural land (1033 ha), forests and pastures (234 ha) take the majority of the district area. Forest land occupies an area of 821 ha, meadows 56 ha, and pastures 21 ha. The forests consisting of oaks and hornbeams are tied only to the steeper hill slopes.

The village of Nenince was first mentioned in 1135. The original name Nena (Nene) comes from the personal name Něna, documented as early as in the 9th century. The name Nenince is formed from this Něna by adding the suffixes *-in* and *-ce*. The name Luka-nenince comes from Dobák's son Luka, who was the village owner. In 1410, two villages are already mentioned (Michalová/Michal 1980, 263).

From the point of view of the vineyard landscape research and reconstruction, there are very interesting old wine cellars on the territory under investigation. Remains of cellars are located 1.8 km northeast of the centre of Nenince village. Original vineyards almost completely disappeared in this area as the terrain was significantly changed by terracing during the construction of new and larger vineyards in the second half of 20th century (Fig. 48). Remains of approximately 21 buried cellars are spread in a horseshoe-like shape on a slightly wooded elevation of the valley, located below the overgrown vineyard. The entrances to the cellars are mostly buried and appear only as convex terrain formations. In some cases, openings of buried entrances can be observed, where barrel vaulting can be seen (Fig. 49).

FIGURE 48

FIGURE 49

In three places, convex formations are arranged directly above each other, indicating a possible vertical arrangement of some entrances. In addition to these objects, the original access road leading from the village is also visible in the terrain. The road splits in front of the cellars, allowing for easy access to each cellar. However, the approximate dating of these cellars is very difficult due to the lack of available sources. It is clear from military maps that the place was already forested at the end of 18th century. Exposed wine cellars are not unusual in this region, similar ones can also be found in locations such as Rykynčice and Malý Horeš. The forest provided some protection to these structures in times of unrest and could also serve as a refuge for local residents.

Source base of digital reconstruction

The wine-growing cultural landscape in this area has undergone significant changes. In particular, changes in the vineyard areas, changes in the way of use in the 20th century caused by mechanization of production as well as the condition of current vineyard landscape, represented by preserved landscape structures, were monitored. For the given goal, the most relevant map

FIGURE 50

materials capturing the territory in question have to be selected. In this case, the source base consisted mainly of maps from the middle of 19th century to the present. For this purpose, the maps of the first and second military mapping were used, faithfully capturing the historical landscape from the end of 18th and during the 19th century (Fig. 50). These maps were compared in QGIS program with military topographic maps in scales 1:25,000 and 1:50,000 from the second half of 20th century, available through the GeoData plug-in in the named program. The maps from the 1950s were chosen for their accuracy in capturing the terrain, which was important given the less than half a century since the reambulation of the third military mapping began. This time gap is not significant for our purposes. To display the current landscape, we used current orthophotographic images available on the ZBGIS portal, which is operated by the Office of Geodesy, Cartography and Cadastre of the Slovak Republic (ÚGKK SR in Slovak). The map data were georeferenced to each other in the QGIS environment. Digitized maps were mostly georeferenced automatically, but in the case of military mapping, manual georeferencing or identifying the vineyard borders according to the terrain shape and configuration was necessary. The georeferencing of

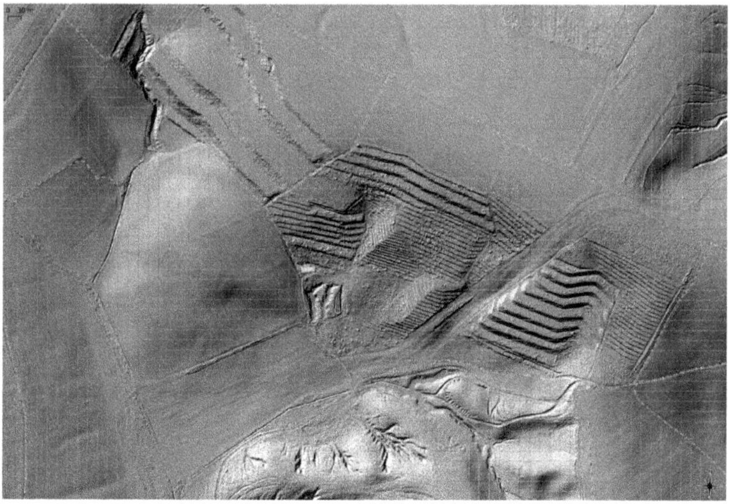

FIGURE 51

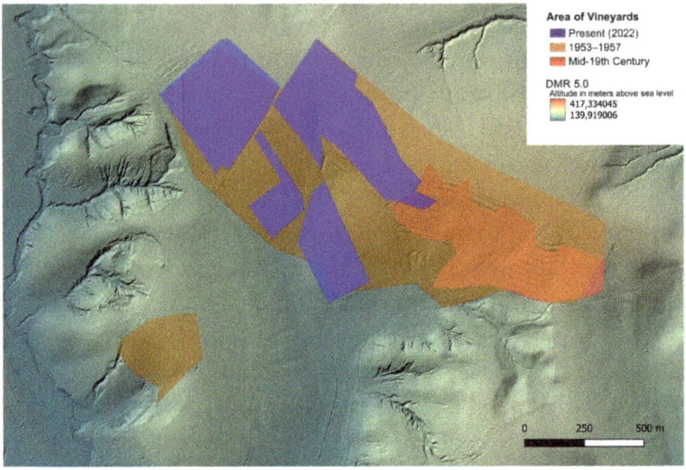

FIGURE 52

current map data was carried out to locate the older map data on the current landscape surface more accurately.

Finally, a DMR 5.0 raster layer was created for the given territory from data downloaded from the ÚGKK server (Fig. 51). On the prepared maps, the vineyard areas changing over time on the selected territory were identified and plotted using polygonal vector layers (Fig. 52). Three vector layers were created that represent the extent of vineyard areas in these periods:

1. First half of the 19th century
2. Years 1953–1957
3. Present

In addition to map sources, old photographs from ethnographic research carried out in the second half of 20th century that were published in book (Drábiková 1989) and online on the Uetnofolk website were selected to trace architectural parts of the vineyard. These documents were significant mainly in the case of manual reconstruction of vineyard houses and cellars. In this case, all image sources had only analogical meaning.

After this process, the created map data were imported into the Blender program.

Reconstruction methods and procedure

The first step was to create a 3D map of the area that could be imported into the Blender program. The map was created in the QGIS program using the DEMto3D plugin, from which the spatial map in .stl format was exported. After importing into the program in Blender, to remove the terraces created by mechanization in the middle of 20th century was necessary. The process took place in Sculpting mode with Flatten and Smooth tools. After the end, the defunct vineyard area was deprived of recent anthropogenic elements.

For terrain texturing, historical maps are applied as textures to 3D landscape model. UV mapping is used to assign accurate textures to individual parts of the model. The textures were prepared in graphic editors (GIMP), where colour and contrast were adjusted to depict the historical landscape appearance as faithfully as possible. Used in this way, they ensure that the terrain model reflects geographical and historical details of the second half of 19th century. In order to depict the original land cover more accurately,

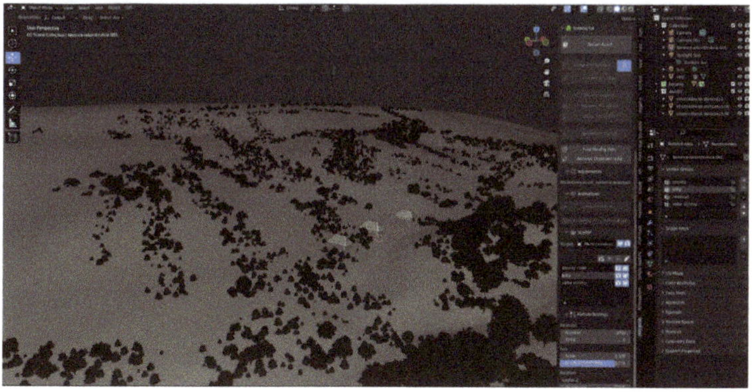

FIGURE 53

our own vegetation maps and road network in the vineyard was drawn in the Bleder program.

With the Botaniq add-on, vegetation was added. Vines and other plant models are selected and placed from the extensive plugin library. When placing the plants, historical accuracy was taken care of, typical spacing between rows of vines are modeled and plant types that were common in the given period are used (Fig. 53). The plant parameters, such as size, density and shape were adjusted to reflect the conditions of the second half of 19th century veraciously.

Completing the model included adding details, such as vineyard houses, roads, fences, and other structures. These elements were modeled by hand or ready-made objects from the Blender libraries were used, which were modified according to historical maps and available sources. Smaller details, such as stones, bushes and other natural elements are also added to emphasize the realism of the model (Fig. 54).

Lighting and rendering are the last steps. Lighting conditions were set to underline the details of terrain and vegetation. They use realistic light and shadow settings to create an authentic atmosphere of a historic vineyard. Rendering is done using the Cycles mode, while taking care of the final image quality and its fidelity to historical background.

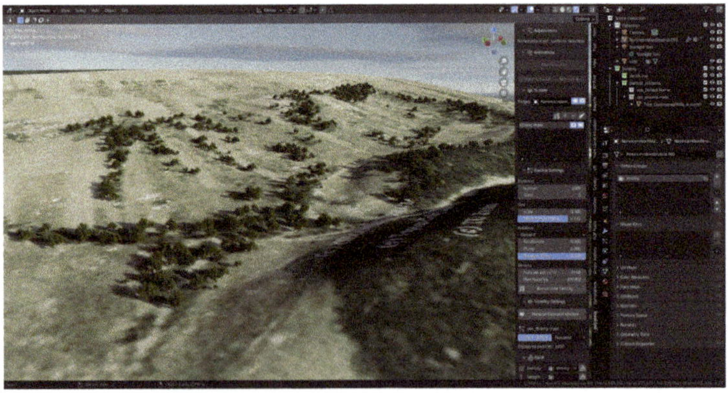

FIGURE 54

Evaluation of uncertainty visualization

For evaluation of the defunct vineyard landscape near Nenince village, one certainty degree evaluation was selected that was aimed in assessing the degree of historical landscape reconstruction.

Changes in geo-relief of the territory: the landscape geo-relief disturbed to a certain extent as a result of mechanization and automation of agricultural production in the second half of 20th century.

Changes in micro-relief of the territory: the micro-relief has undergone significant changes that were caused by the above-mentioned changes in use of agricultural land.

Changes in original landscape cover: The original land cover has undergone changes, especially in the area where the original vineyards were situated. The historical oak and hornbeam forest and pastures are preserved only in fragments.

Research on anthropogenic phenomena: no research known at the time of reconstruction.

Research on environment: no research known at the time of reconstruction

Anthropogenic structures: Anthropogenic structures from the second half of 20th century are significantly preserved, but they are irrelevant for the reconstruction. The wine cellars dug into the slope and preserved in the forest can be considered older structures without further dating.

Historical mapping on a small scale (from 1:1000 to 1:5000 – simplified content, no contour lines, no plans): no small-scaled maps of the territory available in the time of reconstruction.

Historical mapping on a medium scale: three Old-Hungarian military maps have been preserved, where the vineyard borders and several timber-clay constructions are visible.

Historical mapping on a large scale: large-scale maps are irrelevant for the reconstruction of the vineyard, as it is a rather small area.

Historical aerial photographs: digitized aerial orthographic photographs from the 1950s preserved.

Detailed cartographic localization of the territory (LiDAR DTM DMR): data obtained by LIDAR are available in the point cloud or in raster DMR map.

3D digital documentation (drone): for comparison, the park was photographed with UAV device. However, its mapping in the form of perpendicular orthographic images or 3D recording was not done, and therefore we evaluate this attribute with only half the value.

LIM implementation: no analyzes known at the time of reconstruction.

Vedutes or historical drawings: during the reconstruction, no one concerning the area under reconstruction was available.

Historical photographs: no one photograph from the time period under reconstruction was available. Analogous photographs a century younger were used indirectly, but they were not reflected in the visualization relevance evaluation.

Written sources: in the time of reconstruction no written sources available that could help reconstruct the defunct vineyard.

Evaluation of the data entered in Table 11. was as follows. The final score from the table after adding the individual attributes is 500 points

Table 11 Evaluation of uncertainty visualization of the defunct vineyard landscape near Nenince village. Created by D. Bešina.

	100	50	0
Changes in geo-relief		■	
Changes in micro-relief			■
Changes in original landscape cover		■	
Research on anthropogenic phenomena			■
Research on environment			■
Anthropogenic structures		■	
Historical mapping – small scale			■
Historical mapping – medium scale	■		
Historical mapping – large scale			■
Historical aerial photographs	■		
Detailed cartographic location (LiDAR, DTM, DMR)	■		
3D digital documentation (drone)		■	
LIM implementation			■
Vedutes or historical drawings			■
Historical photographs			■
Written sources			■
Total		**500**	

out of total possible value of 1600 points. In percentages terms, the certainty degree of the digital visualization of defunct vineyard landscape in Nenince village cadastre is only 31.3 %.

Final reconstruction

The low reconstruction certainty degree of the vineyard from the mid-19th century is caused primarily by intensive anthropogenic interventions in the second half of 20th century, insufficient state of research and, last but not least, by small number of authentic sources. The reconstruction shows the expected layout of the vineyard plots, which are interspersed with narrow fields or pastures. There is also a vegetation cover of fruit trees that can still be found in some extinct vineyards today (Fig. 55–56). The space where the sunken wine cellars were located was placed in a sparser forest. In the gentle valley through which the original access road passed, we also placed above-ground wine houses to illustrate the vineyard architectural components in a better way. This reconstruction can only serve as a very hypothetical representation of a potential appearance of vineyards in the south of central Slovakia with emphases put on presentation and popularization possibilities.

FIGURE 55

FIGURE 56

As in the case of previous reconstruction examples alike, the certainty level can be increased by further interdisciplinary research that deepens the knowledge in the area under investigation.

In this chapter, specific examples are presented to describe the methods and procedure of evaluating the certainty degree in digital reconstructions of historical monuments, along with possibilities of its practical use. The assessment of the certainty level proved to be crucial for the authenticity and scientific value of digital visualizations.

In reconstruction of Čebovce water mill, an uncertainty classification was used, which determined the relevance of individual parts of the display based on available sources and analogies. The result was a low certainty degree, which affected the reconstruction credibility. Nevertheless, the reconstruction provides valuable visual and research outputs that can be supplemented with new findings in the future.

The reconstruction of the manor house in Dolná Strehová was carried out with a higher certainty degree, as historic photographs, drawings and orthographic images were available. This project highlighted the importance of detailed documentation and digital technologies, such as photogrammetry and UAV, in creating accurate and authentic models. Evaluation of a certainty degree helped identifying areas of higher and lower preservation levels, which is relevant for scientific and public presentation.

In the example of historical vineyard landscape in Nenince, the reconstruction was complicated by significant changes in the landscape during the 20th century. The certainty degree evaluation was focused on changes in geo-relief, micro-relief and original landscape cover. Its results pointed to the need for an interdisciplinary approach and cooperation of different scientific fields in the evaluation of historical landscape features.

The certainty degree evaluation in digital reconstructions provides critical data for experts and the public, thus playing a crucial role in the process of conservation and presentation of historical objects. This evaluation enables a detailed analysis of available sources and identification of authentic elements that are most important for the reconstruction. It also highlights areas where sources are insufficient or unclear, hence identifying the need for further research and verification.

In this way, the certainty degree evaluation can contribute to dynamic updating of digital models based on new archaeological finds, historical documents or technological advances. This flexibility allows the models to be kept up-to-date and accurate, which is crucial for their scientific value and credibility. Transparency in model evaluation and updating fosters trust among experts, historians, and the lay public.

The resulting reconstructions, which take into account the certainty degree, can be effectively used in museum exhibitions, where they provide visitors with a visually rich and informative view of historical objects and landscapes. In academic and research projects, these models can serve a basis for further studies, simulations and education. In addition, they are a valuable tool in public presentations where they can educate and engage the general public on issues of historical and cultural value.

Chapter 11

Accessibility, archiving and presentation

The topic of accessibility, archiving and presentation of digital monuments is very extensive and complex, overlapping with many technical, legal, ethical and social areas. For this reason, it is beyond the scope of this publication to cover all aspects in full breadth and depth. In the following sections, the focus is hence laid on a brief overview of vital theoretical aspects of this topic, drawing on the author's own experience and knowledge. The aim is to provide basic information and stimuli for further study and discussion in this area.

Accessibility of digital content is an integral part of the present-day digital era, being of fundamental importance for cultural, educational and social inclusion. At a time when information and knowledge are constantly growing and evolving, the widespread availability of digital content enables people around the world to access resources that can enrich their lives, provide educational opportunities, and support personal and professional development.

Access to digital content is a key to democratizing knowledge. In the information age, every individual has the right to access information that can influence their decisions and lifestyle. Making digital content widely available means enabling a wider range of individuals, regardless of their geographic location, socio-economic status or education, to engage in the global conversation and use digital resources for their personal and professional enrichment.

Sharing of digital data between experts within various disciplines is critically important for many aspects of scientific progress and effective cooperation. The first and fundamental benefit is the promotion of collaborative research, which allows scientists from different disciplines and geographical areas to combine their skills and perspectives, thereby expanding the scope and depth of their studies. This cooperation leads to a better understanding of the problems investigated and to innovative solutions.

Another cardinal advantage is improved reproducibility of the results. Transparent data sharing allows for independent validation and verification of results, which is essential for ensuring the credibility of scientific findings. At the same time, this supports the ethical integrity of scientific work and strengthens public confidence in the scientific process.

From an economic and practical point of view, data sharing is an effective way to optimize resources. It allows the researchers to use existing data without the need to re-acquire it, thus saving time, money and other resources. This enables faster scientific progress, and at the same time supports wider application and use of the results in practice.

Sharing the data also greatly expands the research possible impact, as freely available data can inspire new discoveries and applications across disciplines. The interdisciplinary use of data opens the door to new perspectives and solutions, which can lead to breakthrough innovations at the intersections of traditional scientific disciplines.

Finally, ethical and social obligations are significant as well. In cases where the data has been obtained from public sources or is related to important social issues, sharing it can be seen as a moral responsibility. Sharing this data can contribute to the public good and help solving global problems, which increases the social and scientific impact of research.

The sharing of digital data by experts is thus of fundamental importance both for the scientific community and the wider society, as it enables faster progress, ethical integrity and wider social contribution of science (Diagram 7).

Accessibility	• Technological accessibility • *Legal accessibility*
Archiving	• *Technical archiving* • *Archiving strategy*
Presentation	• *Latest technologies* • *Adjusting to user´s needs* • *Cooperation of professionals*

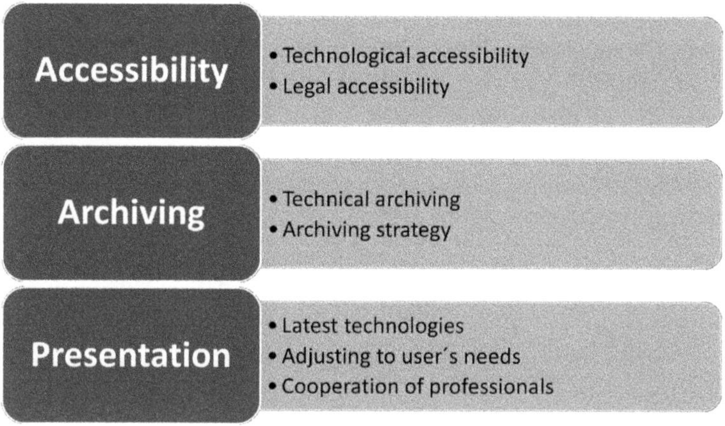

DIAGRAM 7

Wide availability of digital content also encourages innovation and creativity. When people have access to diverse information and ideas from different parts of the world, it can inspire new ideas and collaboration that are transcending the cultural and geographic boundaries. This approach encourages cultural diversity and collaboration, thus opening the door for new technological and artistic discoveries. In addition, the wide availability of digital content can increase transparency and accountability in both the public and private sectors. When people have access to information about the activities of governments and corporations, they can better monitor their activities and demand accountability for decisions that affect the public interest.

Finally, wide availability of digital content is essential in the process of education. Digital learning resources, such as online courses, educational videos, and access to e-libraries, enable students and teachers to access a wide range of educational materials, thus increasing the quality and accessibility of education. At a time when education is considered a fundamental pillar for personal and social development, it is substantial that this digital content be accessible to all. Therefore, the wide availability of digital content is crucial not only for legal justice and equality of access, but also for fostering global innovation, creativity and cultural diversity, all of which are essential elements for a healthy, dynamic and inclusive society.

Considering the constantly evolving technologies and changes in social standards, the approaches that enable effective management of these resources need to be adapted. This chapter focuses on three basic aspects of accessibility, archiving and presentation of digital monuments, which are an integral part of their conservation for future generations.

Accessibility of digital monuments involves technological aspects together with legal and ethical issues. The legal framework has to be flexible to allow for copyright protection, while at the same time supporting open access and sharing of content. In practice, this means balancing the protection of intellectual property with possibilities for public access and use of digital collections. It is also important to resolve technological accessibility so that digital monuments are accessible to all groups of users, including people with disabilities, and to ensure their adaptability to various platforms and devices. The availability of digital monuments can be improved through the use of cloud technologies, which enable centralized data storing with the possibility of being effectively accessed from diverse places and devices. Cloud solutions also contribute to reducing the cost of maintaining physical infrastructure while increasing data security through advanced encryption technologies. In addition, the attention should be paid to simplifying user interfaces so that digital resources are accessible even to less technically capable users.

Archiving is a technical and strategic challenge. Preserving of digital content in a long-term perspective requires development of standards, which can ensure that digital data remain accessible and readable even in the context of continuous technological development. Considering this, robust solutions for regular data migrations to modern media and software format updates need to be implemented. In addition to technical aspects, the effective archiving also involves creating transparent policies and procedures that can regulate digital resources collecting and managing along with their long-term preservation. Effective digital archiving requires applying of the sustainability concept, which includes constant updating and conversion of digital formats. Institutions, such as libraries, museums and archives should implement regular reviews of their archiving procedures and ensure that digital objects are stored in formats that meet current standards and are compatible with modern technologies. An example of such a practice can be creating of multiple copies of digital files that are stored in different geographic locations to minimize the risk of data loss in the event of a technical failure or natural disaster.

Presentation of digital monuments has to be adapted to constantly changing way how people use to access and interact with digital content. The use of modern technologies such as virtual and augmented reality opens up new possibilities for education and public engagement. Interactive and multimedia presentations can significantly enrich the user experience and provide a deeper understanding of the monument cultural and historical context. However, this approach induces the need to ensure that technological innovation is not just a temporary trend, but it actually adds value to the sustainability and educational potential of digital archives.

Presentation of digital monuments can go beyond traditional methods of display and interaction using augmented reality (AR) and virtual reality (VR) technologies. These technologies provide users with the opportunity to experience historical events, art and culture in very authentic and interesting formats. For example, virtual tours allow people from all over the world to walk through historical sites and museums without being physically present there. AR applications can, in turn, "bring to life" historical figures or events directly in the user's environment, thus providing interactive and educational experiences.

Success in accessibility, archiving and presentation of digital monuments depends on close cooperation of IT professionals, archivists, historians, lawyers and educators. This cooperation should be aimed at developing policies that respect ethical standards, legal regulations and technological possibilities. The exchange of knowledge and experience among various groups of users can contribute to the search for new solutions to the challenges posed by digitization.

In this chapter, an overview of the author's own theoretical aspects is presented that is highlighting the technological, legal, ethical and social challenges associated with this topic. A success in this field depends on close collaboration of experts from different disciplines and on creating the policies and solutions that can support the long-term sustainability and accessibility of digital monuments. Technological innovations, legal frameworks and ethical principles have to be constantly updated and adapted to the changing reality in order to ensure that digital heritage is accessible, safe and usable for all.

By focusing on inclusiveness, cooperation and continuous improvement, we can achieve that digital monuments in digital landscape will be a valuable source of knowledge, innovation and cultural identity for all of humanity. With this approach, the history can be preserved in the developing dynamic and inclusive society that values its heritage and uses it for its further growth and progress.

Summary

The publication “Digital Monuments in Digital Landscape” can make a fundamental contribution to the protection, reconstruction and presentation of historical monuments and landscape structures using modern digital technologies. The work offers a comprehensive overview and analysis of how contemporary tools, such as landscape information modeling (LIM) or historic building information modeling (HBIM), can be effectively integrated into the monument care. The book is focused on both the theoretical aspects and methodologies and the practical applications of these technologies in the real world, emphasizing the importance of interdisciplinary cooperation of experts from various scientific fields, such as geography, archaeology, architecture and computer science.

One of the key contributions of the monograph is the design and application of its own evaluation system for virtual reconstructions of monuments and historical landscapes, which brings fundamental progress in the field of digital monument care. This uncertainty classification system is an innovative tool that allows accurate evaluation of various aspects of digital models, which is critical to maintaining their authenticity and historical veracity. The system is designed to provide a structured framework for identifying and quantifying uncertainties that may arise in digital reconstruction, including geo-relief change, preservation of anthropogenic structures, historical mapping, and availability and quality of 3D digital documentation.

This evaluation system is particularly useful in decision-making processes where it is necessary to determine to what extent a digital reconstruction is reliable and how it can be used in practice, whether in the field of research, education or presentation to the public. Using this system, potential risks and limitations associated with the use of digital models can be better assessed and the areas that require further research or improvement can be identified. This approach contributes to improving the quality of digital reconstructions as well as to increasing the transparency and trustworthiness of these models, which is crucial for their wide acceptance by the professional and lay public.

In addition, the evaluation system can be flexibly adapted, which allows it to be adapted to different types of monuments and historical landscapes, making it a universal tool for the monument care in the digital space. Its implementation means a significant step forward for the cultural heritage digital protection, as it enables conserving the historical integrity of reconstructed objects and their adapting to current technological conditions, which can guarantee their long-term sustainability and accessibility. This system therefore not only contributes to scientific research, but also has practical use in conservation projects that aim to preserve cultural heritage for future generations.

The monograph also points out the economic and practical benefits of sharing digital data between experts and the public. This approach promotes more effective management and protection of cultural heritage, while reducing the cost and time needed to implement conservation projects. The use of technologies such as HBIM enables better planning, reducing errors and conflicts in historic building reconstruction projects, thus leading to improved overall heritage management. In the final parts of the monograph, future trends in the field of digital preservation are discussed, while identifying the main challenges associated with the sustainability of digital archives and the need for continuous development of technological tools.

Altogether, the monograph represents an invaluable source of knowledge and methodological procedures that can be widely used in practice - from research and conservation projects to education and public presentation of historical monuments. Offering innovative approaches and emphasizing the interdisciplinary cooperation, it opens up new possibilities for the cultural heritage protection and accessibility in the digital age.

References

3D-ICONS Guidelines [online]. 3D ICONS, 2014 [Accessed on 18.7.2024]. Available at: https://www.scribd.com/document/403960391/3D-ICONS-Guidelines-pdf

ADAMOVE, Norbert. 2015. *Dolná Strehová očami Strehovčana*. Dolná Strehová: Norbert ADAMOVE, 2015. 247 p. ISBN 978-80-9720338-8-7.

AHMAD, Ahmad / ALIYU, Abdullahi Adamu. 2012. *The Need for Landscape Information Modelling (LIM) in Lanscape Architecture* [online]. Conference: 13th Digital Landscape Architecture Conference [Accessed on 18.3.2024]. Available at: https://www.researchgate.net/publication/230857069_The_Need_for_Landscape_Information_Modelling_LIM_in_Lanscape_Architecture

AKKER, van den Chiel. 2011. *Digital hermeneutics: Agora and the online understanding of cultural heritage* [online]. WebSci '11: Proceedings of the 3rd International Web Science Conference, 2011, vol. 10, p. 1–7 [Accessed on 12.7.2024]. Available at: https://dl.acm.org/doi/10.1145/2527031.2527039

ALBERTY, J. 1989. *Novohrad. Dejiny 2/1*. Lučenec: Vydavateľstvo Osveta, 1989. 456 p.

ANTONOPOULOU, Sofia / BRYAN, Paul. 2017. *BIM for Heritage: Developing a Historic Building Information Model* [online] Swindon: Historic England. 78 p. [Accessed on 10.3.2024]. Available at: https://historicengland.org.uk/advice/technical-advice/recording-heritage/

APOLLONIO, Fabrizio. 2015. Classification schemes and model validation of 3D digital reconstructionprocess [online]. In *Proceedings of the 20th*

International Conference on Cultural Heritage and New Technologies. New York: AMC Press, 2015. [Accessed on: 18.7.2024]. Available at: http://www.chnt.at/wp-content/uploads/eBook_CHNT20_Apollonio_2015.pdf

APOLLONIO, Fabrizio. 2016. Classification Schemes for Visualization of Uncertainty in Digital Hypothetical Reconstruction [online]. In *3D Research Challenges in Cultural Heritage II: How to Manage Data and Knowledge Related to Interpretative Digital 3D Reconstructions of Cultural Heritage*, eds. S. Münster, M. Pfarr-Harfst, P. Kuroczynski and M. Ioannides, Cham: Springer International Publishing, 2016, p. 173–198 [Accessed on: 24.7.2024]. Available at: https://chnt.at/wp-content/uploads/eBook_CHNT20_Apollonio_2016.pdf

APOLLONIO, Fabrizio / GIOVANNINI, Elisabetta Giovannini. 2015. A paradata documentation methodology for the Uncertainty Visualization in digital reconstruction of CH Artifacts [online]. In *ScIentific Research and Information Technology*, vol. 5 is.1, p. 1–24. [Accessed on: 3.8.2024]. Available at: DOI: 10.2423/i22394303v5n1p1

ARAYICI, Y., Counsell, J., & Mahdjoubi, L. (2017). *Heritage Building Information Modelling*. London: Routledge.

ARGYROU, Argyro. / AGAPIOU, Atos. 2022. Review of Artificial Intelligence and Remote Sensing for Archaeological Research [online]. In *Remote Sens*. vol. 14, 2022. p. 1–23 [Accessed on: 19.7.2024]. Available at: https://www.researchgate.net/publication/365790779_A_Review_of_Artificial_Intelligence_and_Remote_Sensing_for_Archaeological

BAGNOLO, Vincenzo et al. 2019. HBIM for archaeological sites: from SfB based survey to algorithmic modeling [online]. In *The International Archives of the Photogrammetry, Remote Sensing and Spatial Information Sciences*, 8th Intl. Workshop 3D-ARCH "3D Virtual Reconstruction and Visualization of Complex Architectures", roč. XLII-2/W9, 2019, Bergamo: Italy. s. 57–63 [Accessed on: 18.3.2024]. Available at: https://isprs-archives.copernicus.org/articles/XLII-2-W9/57/2019/isprs-archives-XLII-2-W9-57-2019.html

BAKKER, Geeske / MEULENBERG, F / RODE, de Jan. 2003. Truth and credibility as a double ambition: reconstruction of the built past, experiences and dilemmas [online]. In *The Journal of Visualization and Computer Animation*, vol. 14, is. 3, 2003, p. 159–167 [Accessed on: 18.7.2024]. Available at: https://www.researchgate.net/publication/220357291_Truth_

and_credibility_as_a_double_ambition_Reconstruction_of_the_built_past_experiences_and_dilemmas
BANFI, Fabrizio / BARAZZETTI, Luigi / PREVITALI, M. / RONCORONI, F. 2017. Historic BIM: *A New Repository for Structural Health Monitoring* [online]. Arch. Photogramm. Remote Sens. Spatial Inf. Sci., XLII-5/W1, 269–274 [Accessed on: 13.3.2024]. Available at: https://isprs-archives.copernicus.org/articles/XLII-5-W1/269/2017/
BARCELÓ, J. A. 2006. Automatic Archaeology. Bridging the gap between Virtual Reality, Artificial Intelligence and Archaeology [online]. In *Theorizing Digital Cultural Heritage: A Critical Discourse (Media in Transition)* Edited by Fiona Cameron (Editor), Sarah Kenderdine. Cambridge (MA), The MIT Press, 2006 ISBN 0262033534 [Accessed on: 18.7.2024]. Available at: https://www.researchgate.net/publication/225469871_Visual_Analysis_in_Archaeology_An_Artificial_Intelligence_Approach
BARKER, Philip. 1982. *Techniques of Archaeological Excavation*. London: B.T. Batsford Ltd., 1982. 281 p. ISBN 0-7134-7169-7.
BARKER, Philip. 1986. *Understanding Archaeological Excavation*. London: B.T. Batsford Ltd., 1986. 189 p. ISBN 0-7134-3631-X.
BARNHART K., Richard / HOTTMAN B., Stephen / MARSHALL M., Douglas / SHAPPEE, Eric. 2012. *Introduction to Unmanned Aircraft Systems*. CRC Press: Taylor & Francis Group. 2012. 214 p. ISBN 978-1-4398-3521-0.
BARRILE, Vincenzo / BERNARDO, Ernesto / BILLOTA, Giuliana. 2022. An Experimental HBIM Processing: Innovative Tool for 3D Model Reconstruction of Morpho-Typological Phases for the Cultural Heritage. In *Remote Sens*, vol. 14, no. 5, 2022. p. 1–17. ISSN ISSN: 2072-4292.
BARTOŠÍKOVÁ, Tereza. 2023. Vodné mlyny v Zbierke listín k vodnej knihe
BARČÁKOVÁ, Ivana. 2001. Prístupy k hodnoteniu estetickej (vizuálnej) kvality krajiny. In *Geografický časopis*, 2001, vol. 53., no. 4. p. 343–345. ISSN 2453-8787.
BELJAK, Ján / MITÁŠ, Vladimír. 2013. Praveké a stredoveké nálezy z Dolnej Strehovej. In *AVANS 2009.* Nitra, Archeologický ústav SAV Nitra 2013. p. 40–42. ISBN 978-80-89315-54-3.
BENVENUTO, Francesco et al. 2024. *Augmented, Mixed and Virtual Reality* [online]. Engineering [Accessed on: 17.7.2024]. Available at: https://www.eng.it/resources/whitepaper/doc/ar-mr-vr/ar-mr-vr-whitepaper-en1.pdf

BENČO, Jozef. 2001. *Metodológia vedeckému výskumu*. Bratislava: Vydavateľstvo IRIS, 2001. 194 p. ISBN 80-89018-27-0.

BOLTIŽIAR, Martin / OLAH, Branislav. 2009. *Krajina a jej štruktúra (Mapovanie, zmeny a hodnotenie).* Nitra: Fakulta prírodných vied UKF v Nitre, 2009. 148 p. ISBN 978-80-8094-552-7.

BONISSONE Piero Patrone / TONG, Richard M. 1985. Editorial: Reasoning with uncertainty in expert systems. In *International Journal of Man-Machine Studies*, vol. 22, no. 3. p. 241–250.

BORKOWSKI, Andrzej / WYSZOMIRSKI, Michał. 2021. Landscape Information modelling: an important aspect of BIM modelling, examples of cubature, infrastructure, and planning projects [online]. In *Geomatics Landmanagement and Landscape*, 2021. s. 7–22 [Accessed on: 19.3.2024]. Available at: https://www.researchgate.net/publication/351853042_Landscape_Information_modelling_an_important_aspect_of_BIM_modelling_examples_of_cubature_infrastructure_and_planning_projects

BOROVSZKY, Samu. 1911. *Magyarország vármegyéi és városai*. Nógrádvármegye. Országos Monográfia Tarsaság, Budapest. 748 p.

BOTÍK, Ján / SLAVKOVSKÝ, Peter. 1995. *Encyklopédia ľudovej kultúry Slovenska 1*. Bratislava: VEDA, Vydavateľstvo SAV. 484 p. ISBN 80-224-0234-6.

BREJCHA, M. et al., 2015. *Metodika digitalizace, 3D dokumentace a 3D vizualizace jednotlivých typů památek*. Národní památkový ústav, Ústí nad Labem 2015. 60 p. ISBN 978-80-7414-954-2.

BRUSAPORCI, Stefano. 2017. The Importance of Being Honest: Issues of Transparency in Visualization of Architectural Heritage [online]. In IPPOLITO, A. ed. *Handbook of Research on Emerging for Architectural and Archaeological Heritage*, 2017, p. 66–92. Hershey: IGI Global [Accessed on: 21.7.2024]. Available at: https://www.researchgate.net/publication/307637893_The_Importance_of_Being_Honest_Issues_of_Transparency_in_Digital_Visualization_of_Architectural_Heritage

BRŮNA, V. et al., 2015. *Metodika 3D dokumentace a vizualizace interiérů památkových objektů* [online]. Národní úložiště šedé literatury, 2015 [Accessed on: 26.7.2024]. Available at: https://invenio.nusl.cz/record/253869/files/nusl-253869_1.pdf

BUDAY, Peter. 2017. K dejinám pamiatkovej ochrany na Slovensku v rokoch 1846–1919. In *Pramene k umelecko-historickému bádaniu a ochrane pamiatok na Slovensku (1846–1918)*. Bratislava: Stimul, 2017. p. 8–24. ISBN 978-80-8127-198-4.

BÖNDE GOGOVÁ, Stanislava. 2022. Pominuteľné a večné. O prezentácii kultúrneho dedičstva a kresťanskej tradícii tribečského spoločenstva [online]. In *Studia Aloisiana*, vol. 13, no. 1., Teologická fakulta, Trnavská univerzita v Trnave [Accessed on: 12.8.2024]. Available at: https://www.tftu.sk/sites/default/files/sa-13-2022-1-vnutro.pdf

bývalej Abovsko-turnianskej župy v Štátnom archíve Košice. In BARTOŠÍKOVÁ, Tereza et al. *K dejinám mlynov/výskum a mapovanie*. Pamiatkový úrad Slovenskej republiky, Vodné mlyny Oblazy, n.o. Zborník príspevkov zo seminára Zborník príspevkov zo seminára 14. 12. 2021, 2023. p. 177–128. ISBN 978-80-89175-97-0.

CARTER, Jamie / SCHMIDT, Keil / WATERS, Kirk et al. 2012. *An Introduction to Lidar Technology, Data, and Applications* [online]. National Oceanic and Atmospheric Administration (NOAA) Coastal Services Center, 2012. [Accessed on: 7.3.2024]. Available at: https://coast.noaa.gov/data/digitalcoast/pdf/lidar-101.pdf

CASTELLANO-ROMÁN, Manuel / PINTO, Francisco. 2019. *Dimensions and Levels of Knowledge in Heritage Building Information Modelling, HBIM: The model of the Charterhouse of Jerez (Cádiz, Spain)* [online]. Digital Applications in Archaeology and Cultural Heritage. 14 [Accessed on: 18.3.2024]. Available at: https://www.researchgate.net/publication/335457317_Dimensions_and_Levels_of_Knowledge_in_Heritage_Building_Information_Modelling_HBIM_The_model_of_the_Charterhouse_of_Jerez_Cadiz_Spain

CHAMPION, Malcolm Erik. 2021. *Virtual Heritage*. London: ubiquity press, 2021. p. 45. ISBN 978-1-914481-00-0.

CHAPINAL-HERAS, Diego / DÍAZ-SÁNCHEZ, Carlos. 2024. A review of AI applications in human sciences research [online]. In *Digital Applications in Archaeology and Cultural Heritage*, vol. 32, 2024. p. 1–6 [Accessed on: 18.7.2024]. Available at: https://www.sciencedirect.com/science/article/pii/S2212054823000334

CHRASTINA, Peter. 2001. Krajina ako jeden zo styčných fenoménov prírodných a spoločenských vied. In *Acta Historica Nitriensiae*, vol. 4, Nitra: Filozofická fakulta UKF, p. 334–344.

CHRASTINA, Peter. 2005. Historická geografia na Slovensku: minulosť, súčasnosť a perspektívy. In *Historická geografie*, vol. 33, 2005. p. 420–432. ISBN:80-7286-080-1.

CHRASTINA, Peter. 2010. Výskum krajiny (z aspektu historickej geografie a krajinnej archeológie). In *Folia geographica*, vol. 40, no. 16, p. 125–142. ISSN 1336-6157.

CHRASTINA, Peter. 2023. *Príbehy krajín minulosti*. Trnava: Univerzita sv. Cyrila a Metoda v Trnave, 2023. p. 151. ISBN 978-80-572-0336-0.

CIANCI, Maria Grazia / MOLINARI, Matteo. 2019. Information modeling and landscape: intervention methodology for reading complex systems [online]. The International Archives of the Photogrammetry, Remote Sensing and Spatial Information Sciences, 8th Intl. Workshop 3D-ARCH "*3D Virtual Reconstruction and Visualization of Complex Architectures*", 6–8 February, Bergamo, vol. XLII-2/W9, 2019 [Accessed on: 18.3.2024]. Available at: https://pdfs.semanticscholar.org/f246/1e02f6fa5db7613de26674ef4c2505ec5a5a.pdf

COBB, Peter, / LARGE, J. 2023. Language Models and Generative AI, Oh My! Archaeology in the Time of ChatGPT, Midjourney, and Beyond [online]. In *Advances in Archaeological Practice*, vol. 11, no. 3, 2023, p. 363–369 [Accessed on: 19.7.2024]. Available at: https://www.researchgate.net/publication/374090951_Large_Language_Models_and_Generative_AI_Oh_My_Archaeology_in_the_Time_of_ChatGPT_Midjourney_and_Beyond

DADRAS, Mohsen et al. 2015. Spatio-temporal analysis of urban growth from remote sensing data in Bandar Abbas city, Iran [online]. In *The Egyptian Journal of Remote Sensing and Space Sciences*, vol. 18, 2015. p. 35–52 [Accessed on: 18.7.2024]. Available at: https://www.researchgate.net/publication/275152884_Spatio-temporal_Analysis_of_Urban_Growth_from_Remote_Sensing_Data_in_Bandar_Abbas_City_Iran

DAUGSTAD, Karoline / GRYTLI, Eir. 1999. How to study and manage a multihistoric landscape. In *Norsk geografisk Tidsskrift*, 1999. vol. 53. p. 85–92. ISSN 0029-1951.

DENARD, Hugh. et al. 2012. *The London Charter for the Computer-Based Visualisation of Cultural Heritage* [online]. 2012, [Accessed on: 23.10.2021]. Available at: https://www.londoncharter.org/fileadmin/templates/main/docs/london_charter_2_1_en.pdf

DINKOV, Davis / VATSEVA, Rumiana. 2016. 3D modeling nad Visualization for Landscape simulation [online]. In *Proceedings, 6th International Conference on Cartography and GIS*, 13–17 June 2016, Albena, Bulgaria, 2016. p. 320–333. [Accessed on: 17.7.2024]. Available at: https://cartography-gis.com/docsbca/iccgis2016/ICCGIS2016–33.pdf

Dohovor o ochrane architektonického dedičstva [online]. Štrasburg, Rada Európy, oddelenie publikácie dokumentov, ISBN 92-871 -0799-8 Február 1990 [Accessed on: 6.2.2024]. Available at: https://www.culture.gov.sk/wp-content/uploads/2020/02/DOHOVOR_O_OCHRANE_ARCHITEKTONICKeHO_DEDIcSTVA_EURoPY.pdf

Dohovor o ochrane svetového kultúrneho a prírodného dedičstva [online]. [Accessed on: 6.2.2024]. Available at: https://www.pamiatky.sk/content/data/File/pamiatkovy_urad/evidencia_pamiatok/SKD/dohovor_skd.pdf

DORE, Conor / MURPHY, Maurice. 2012. Structural Simulations and Conservation Analysis - Historic Building Information Model (HBIM). *The International Archives of the Photogrammetry, Remote Sensing and Spatial Information Sciences*, [online] 40(5/W4), 2012 [Accessed on: 13.3.2024]. p. 351–357. Available at: https://www.researchgate.net/publication/276507176_Structural_Simulations_and_Conservation_Analysis_-Historic_Building_Information_Model_HBIM

DRÁBIKOVÁ, Ema. 1989. *Človek vo vinici*, Bratislava: VEDA Vydavateľstvo Slovenskej akadémie vied, 1989. 260 p. ISBN 80-224-0012-2.

DVOŘÁK, Max. 2004. *Katechizmus památkové péče*. Praha: Národní památkový ústav, 2004. 195 p. ISBN 80-86234-55-X.

DVOŘÁKOVÁ, Viera / HUSOVSKÁ, Ľudmila (eds.). 2002. *Ochrana kultúrneho dedičstva v medzinárodných dokumentoch ICOMOS. 1. diel/ Charty a odporúčania*. Bratislava: ICOMOS Slovensko, 2002. 112 p. ISBN 978-80-88855- 45-4.

DVOŘÁKOVÁ, Viera / HUSOVSKÁ, Ľudmila (eds.). 2003. Ochrana kultúrneho dedičstva v medzinárodných dokumentoch ICOMOS. Charty, smernice a rezolúcie. Bratislava: ICOMOS Slovensko, 2003. 64 s.

DVOŘÁKOVÁ, Viera / HUSOVSKÁ, Ľudmila (eds.). 2004. *Ochrana kultúrneho dedičstva v medzinárodných dokumentoch ICOMOS. 2. diel/Deklarácie.* Bratislava: ICOMOS Slovensko, 2004. 116 p. ISBN 978-80-88855-56-9.

DVOŘÁKOVÁ. Viera 2012. *Kultúrna krajina a jej ochrana z hľadiska kultúrneho dedičstva.* [online]. Košice: Občianske združenie UzemnePlany.sk, 2012. [Accessed on: 13.8.2024]. Available at: https://www.uzemneplany.sk/sutaz/kulturna-krajina-a-jej-ochrana-z-hladiska-kulturneho-dedicstva

El-HAKIM, Sabry F. et al. 2006. On the Digital Reconstruction and Interactive Presentation of Heritage Sites through Time [online]. In IOANNIDES, M., ARNOLD, D., NICCOLUCCI, F., MANIA, K. (eds.) *The 7th International Symposium on Virtual Reality, Archaeology and Intelligent Cultural Heritage VAST*, 2006 [Accessed on: 8.4.2024]. Available at: https://www.researchgate.net/publication/268687110_On_the_Digital_Reconstruction_and_Interactive_Presentation_of_Heritage_Sites_through_Time/citations

Elektronické zdroje:

ERVIN, Stephen. 2023. Landscape Meta-Modeling [online]. [Accessed on: 19.3.2024]. Available at: https://www.researchgate.net/profile/Stephen-Ervin-2/publication/265422803_Landscape_Meta-Modeling/links/54a6a4340cf257a6360a8707/Landscape-Meta-Modeling.pdf

EURÓPSKA KOMISIA. 2019. *Evaluation of the CAP measures applicable to the wine sector - Executive summary.* Luxemburg: EU. 7 p. ISBN 978-92-9-92558-0.

Európsky dohovor o krajine (2000). [online]. SLOV-LEX právny informačný portál. [Accessed on: 10.8.2024]. Available at: https://www.slov-lex.sk/static/pdf/2005/515/ZZ_2005_515_20051125.pdf

EVINK, Eddo - ROMELE, A. 2020. Digital hermeneutics [online]. In *Phenomenological Reviews*, vol. 6, no. 1, 2020. p. 1–9. [Accessed on: 17.7.2024]. Available at: https://www.researchgate.net/publication/361598070_A_Romele_Digital_hermeneutics

FADLI, F., & AlSaeed, M. (2019). Digital heritage preservation using BIM. *Journal of Cultural Heritage*, 40, 12–20.

FAI, S. / GRAHAM, K. / DUCKWORTH, T. / WOOD, N. / ATTAR, R. 2011. Building Information Modelling and Heritage Documentation [online]. In *Proceedings of the XXIII CIPA International Symposium*, Prague, Czech Republic, 12th–16th September [Accessed on: 13.3.2024]. Available at: https://www.researchgate.net/publication/308634697_Building_Information_Modeling_and_Heritage_Documentation

FALŤAN, Vladimír / OŤAHEL, Ján / GÁBOR, Marián / RUŽEK, Ivan. 2018. *Metódy výskumu krajinnej pokrývky*. Vysokoškolská účebnica. Bratislava: Univerzita Komenského v Bratislave, 2018. 124 p. ISBN 978-80-223-4441-8.

FELCANOVÁ, Zuzana / FELCAN, Michal / LIESKOVSÝ, Tibor. 2021. *Archeológia neviditeľného. Prípady nedeštruktívneho výskumu v archeológii*. Bratislava: Chronos, 2021. 161 p. ISBN 978-80-89027-54-5.

FITRIA, Tira Nur. 2023. Augmented Reality (AR) and Virtual Reality (VR) Technology in Education: Media of Teaching and Learning: A Review [online]. In *International Journal of Computer and Information System (IJCIS)*, vol. 04, no. 01, 2023 [Accessed on: 17.7.2024]. Available at: https://www.researchgate.net/publication/368655016_Augmented_Reality_AR_and_Virtual_Reality_VR_Technology_in_Education_Media_of_Teaching_and_Learning_A_Review

FLAMARY, Rémi et al. 2015. Analysis of Multitemporal Classification Techniques for Forecasting Image Time Series [online]. In *IEEE Geoscience and Remote Sensing Letters*, vol. 12, no. 5, 2015. p. 953–957 [Accessed on: 17.7.2024]. Available at: https://www.researchgate.net/publication/272391174_Analysis_of_Multitemporal_Classification_Techniques_for_Forecasting_Image_Time_Series

FORTE, Maurizio. 2006. A digital "cyber" protocol for the reconstruction of the archaeological landscape: virtual reality and mindscapes. In BALTSAWAS et al. (eds). *Recording, Modeling and Visualization of Cultural Heritage*, London: Taylor & Francis Group, p. 76–85. ISBN 0 415 39208 X.

FORTE, Maurizio. 2007. Ecological Cybernetics, Virtual Reality and Virtual Heritage. In *Theorizing Digital Cultural Heritage. A Critical Discourse*. Cambridge: MIT Press, p. 389 – 407. ISBN 978-02-62033-53-4.

FORTE, Michael. 2008. *Cyber-archaeology: An eco-approach to the virtual reconstruction of the past. Digital Heritage* [online]. 2008 [Accessed on: 21.2.2022] Available at: https://www.researchgate.net/publication/264235162_CYBER-ARCHAEOLOGY_AN_ECO-APPROACH_TO_THE_VIRTUAL_RECONSTRUCTION_OF_THE_PAST

FRIDRICH, Jan / KUBEČKA, Karel. 2014. *Bim Management: The Use of GDL Language for Effective Work in BIM* [online]. Advanced Materials

Research, s. 865–869, 2014 [Accessed on: 18.3.2024]. Available at: https://www.researchgate.net/publication/280957638_Bim_Management_The_Use_of_GDL_Language_for_Effective_Work_in_BIM

FURMÁNEK, Václav. 2000. Slovansko-avarské pohrebisko v Dolnej Strehovej. In *AVANS 1998*. Nitra, Archeologický ústav SAV Nitra 2000. p. 65–66.

GERSHON, Nahun. 1998. Visualization of an Imperfect World [online]. In *EEE Computer Graphics and Applications*, vol. 18, is. 4, 1998. p. 43–45 [Accessed on: 30.7.2024]. Available at: https://users.cs.utah.edu/~miriah/uncertainty/Gershon1998.pdf

GIRI, Chandra P. 2012. *Remote Sensing of Land Use and Land Cover. Principles and Applications*. Boca Raton: CRC Press, 2012. 397 p. ISBN 978-1-4200-7075-0.

GOGOVÁ, Stanislava. 2011. *Archeológia a múzejná prezentácia*. Nitra: Univerzita Konštantína Filozofa v Nitre, 2011. 172 p. ISBN 978-80-80-8094-850-4.

GOGOVÁ, Stanislava. 2014. *Kultúrne a prírodné dedičstvo. Ochrana a legislatíva v minulosti a v súčasnosti. - 1. vyd.* - Nitra: UKF, 2014. cd - ISBN 978-80-558-0547-4.

GOGOVÁ, Stanislava. 2014. Výskum základov stredovekého kostola a cintorína v Krásne nedeštruktívnymi metódami. In *Študijné zvesti AÚ SAV v Nitre*, vol. 55, Archeologický ústav SAV v Nitre, 2014. p. 131–140. ISSN 0560-2793.

GOGOVÁ, Stanislava. / BEŠINA, Daniel. / WIEDERMANN, Egon. 2020. *Archeomuzeológia na prahu nového milénia*. Nitra: Univerzita Konštantína Filozofa v Nitre, 2020. p. 112. ISBN 978-80-558-1532-9.

GOJDA, Martin. 2000. *Archeologie krajiny*. Praha: Academia, 2000. 238 p. ISBN 80-200-0780-6.

GREENWOOD, Faine. 2015. *How to make maps whith drones* [online]. Chapter 4. Drones and Aerial Observation, 2015 [Accessed on: 5.3.2024]. Available at: http://drones.newamerica.org/primer/Chapter%204.pdf

GREGOROVÁ, Bohuslava / HRONČEK, Pavel. 2022. *Historická geografia. Antropogénne transformácie a využívanie vodných tokov v doline Hrona východne od Banskej Bystrice*. Banská Bystrica: Katedra geografie a geológie FPV UMB, 141 p. ISBN 978-80-557-1995-5.

GRIETHE, Henning / SCHUMANN, Heidrun. 2006. The Visualization of Uncertain Data: Methods and problems [online]. In *Simulation und Visualisierung (SimVis 2006)*, eds. SHUIZE, T.- HORTON, G. - PREIM, B.

and SCHLECHTWEG, S. Magdeburg: SCS Publishing House. 2006, p. 143–156 [Accessed on: 18.7.2024]. Available at: https://www.researchgate.net/publication/216756553_The_Visualization_of_Uncertain_Data_Methods_and_Problems

GÁSPÁR, Albert. 2016. *3D modeling in GIS. Lecture notes on the masters course ITM-E3DM E at the Eötvös Loránd University/Faculty of Informatics* [online]. Eötvös Loránd University (ELTE), Department of Cartography and Geoinformatics, Budapest, 2016. 115 s. [Accessed on: 17.7.2024]. Available at: https://www.inf.elte.hu/dstore/document/362/AlbertG_3D%20modelling%20in%20GIS%20-%20course%20material_v4.pdf

GÜNTHEROVÁ, Alžbeta a kol. 1967. *Súpis pamiatok na Slovensku. Zväzok prvý A/J.* Bratislava: Obzor, 1967. 535 p.

GÜNTHEROVÁ, Alžbeta a kol. 1968. *Súpis pamiatok na Slovensku. Zväzok druhý K/P.* Bratislava: Obzor, 1968. 585 p.

GÜNTHEROVÁ, Alžbeta. 1969. *Súpis pamiatok na Slovensku. Zväzok tretí R/Ž.* Bratislava: Obzor, 1969. 566 p.

GŁOWIENKA, Ewa. 2017. *4D Reconstruction and Visualisation of Krakow Fortress* [online]. 2017 Baltic Geodetic Congress (BGC Geomatics), Gdansk, Poland, 2017, p. 1–5 [Accessed on: 8.4.2024]. Available at: https://ieeexplore.ieee.org/abstract/document/8071435

HADJIMITSIS, Diofantos / THEMISTOCLEOUS, Kyriacos / CUCA, Branka a kol. 2020. *Remote Sensing for Archaeology and Cultural Landscapes Best Practices and Perspectives Across Europe and the Middle East: Best Practices and Perspectives Across Europe and the Middle East.* Cham: Springer Nature Switzerland, 2020. 281 p. ISBN 978-3-030-10978-3.

HALLER, Elisabeta Antonia. 2010. Geospatial Analysis Framework [online]. In *BRAIN. Broad Research in Artificial Intelligence and Neuroscience*, vol. 1, no. 2, 2010. p. 166–171. ISSN 2067-3957 [Accessed on: 17.7.2024]. Available at: https://www.researchgate.net/publication/49595387_Geospatial_Analysis_Framework

HAM, Namhyuk –MOON, Sungkon –KIM, Ju-Hyung / KIM, Jae-Jun. 2020. Optimal BIM staffing in construction projects using a queueing model [online]. In *Automation in Construction*, vol. 113, 2020 [Accessed on: 18.3.2024]. Available at: https://www.sciencedirect.com/science/article/abs/pii/S0926580519301761

HAO, Pu. 2019. *Spatial Analysis* [online]. [Accessed on: 17.7.2024]. Available at: https://www.researchgate.net/publication/332425310_Spatial_Analysis
HARDIN, Brad / McCOOLE, Dave. 2016. *BIM and Construction Managment. Proven tools, Methods and Workflows*. Indianaoplis: SYBEX A Wiley Brand, 2016. 404 p. ISBN ISBN: 978-1-118-94276-5.
HAVRLANT, Miroslav / BUZEK, Ladislav. 1985. *Nauka o krajině a péče o životní prostředí*. Praha: Státní pedagogické nakladatelství, 1985. 126 p.
HERMON, Sorin / NIKODEM Joanna. 2008. 3D Modelling as a Scientific Research Tool in Archaeology [online]. In eds. POSLUSCHNY, A. - LAMBERS, K. And HERZOG, I.. *Layers of Perception. Proceedings of the 35th International Conference on Computer Applications and Quantitative Methods in Archaeology* (CAA), Berlin, Germany, April 2–6, 2007, vol. 10, p. 1-6 [Accessed on: 18.7.2024]. Available at: https://www.academia.edu/1226903/3D_Modelling_as_a_Scientific_Research_Tool_in_Archaeology
Historická ortofotomapa SR: mapy.tuzvo.sk [online]. Laboratórium geoinformatiky, TU Zvolen, 2017–2022 [Accessed on: 30. 7. 2024]. Available at: https:// mapy.tuzvo.sk/hofm/
HLOŽEK, Martin. 2008. *Encyklopedie moderních metod v archeologii. Archeometrie*. Praha: Nakladatelství Libri, 2008. 230 p. ISBN 978-80-7277-230-8.
HOEF, Jay M. Ver. 2001. *Using ArcGIS geostatistical analyst*. New York: Copyright © 2001 ESRI. p. 307.
HOLLÝ, Karol / HRONČEK, Pavel. 2017. Krajina, príroda a životné prostredie v minulosti. In *Forum Historiae*, vol. 11, no. 1, p. 1–7. ISSN 1337-6861.
HORÁČEK, Martin. et al. 2019. *Teoretické základy památkové péče na prahu 21. století*. Praha: Artefactum, nakladatelství Ústavu dějin umění AV ČR, v.v.i. p. 188. ISBN 978-80-88283-27-0.
HRONČEK, Pavel / CHRASTINA, Peter. 2018. Poznámky ku konceptualizácii v environmentálnych dejinách na Slovensku. In *Studia Historica Nitrensia*, vol. 22, no. 2,. p. 408–433. ISSN ISSN 1338-7219.
HRONČEK, Pavel / GREGOROVÁ, Bohuslava / HERČKO, Ivan. 2021. *Hnedouhoľné baníctvo v okolí Veľkého Krtíša*. Banská Bystrica: Novohradský historický spolok, Amtheon, s.r.o. UMB PF, 2021. p. 500. ISBN 978-80-973940-0-4.
HUGH, G. Gauch, Jr. 2012. *Scientific Method in Brief*. New York: Cambridge University Press, 2012. 308 p. ISBN ISBN 978-1-107-66672-6.

Hungaricana Közgyűjteményi portál [online]. Arcanum, Library of Parliament, Hungary [Accessed on: 22.7.2024]. Available at: https://www.hungaricana.hu/en/

HUNKA, Ján / NEVIZÁNSKÝ, Gabriel / OŽĎÁNI, Ondrej. 2001. Geodetické zameranie stredovekého hrádku v Malých Zlievcach. In *AVANS 2001*. Nitra: AÚ SAV v Nitre, 2002. p. 74–76. ISBN 80-88709-59-8.

HÁJEK, Tomáš. 2005. *Zánik a vznik památkových péčí. Filozofie památkové péče*. Praha: Nakladatelství Epocha, 2005. p. 200. ISBN 80-86328-71-6.

INTERRANTE, Victoria / HÖLLERER, Tobias / LÉCUYER, Anatole. 2018. Virtual and Augmented Reality [online]. In *IEEE Computer Graphics and Applications*, vol. 38, no. 3, 2018. p. 28–30 [Accessed on: 17.7.2024]. Available at: https://www.researchgate.net/publication/324493337_Virtual_and_Augmented_Reality

Introduction to Modern Photogrammetry. Wiley, 2001. 496 p. ISBN 978-0471309246.

IT-Empfehlungen für den nachhaltigen Umgang mit digitalen Daten in den Altertumswissenschaften [online]. RWTH Aachen University, 2017 [Accessed on: 3.7.2024]. Available at: https://publications.rwth-aachen.de/record/692137

JANKOVIČ, Vendelín. 1973. Dejiny pamiatkovej starostlivosti na Slovensku v rokoch 1850–1950. In *Monumentorum tutela: Ochrana pamiatok 10*. Bratislava: Pallas, 1973. p. 5–80.

JANKOVIČ, Vendelín. 1981. Dejiny Slovenského ústavu pamiatkovej starostlivosti a ochrany prírody (úsek pamiatkovej starostlivosti) v rokoch 1951–1981. In *Jubilejný zborník: 30 rokov Slovenského ústavu pamiatkovej starostlivosti a ochrany prírody*. Bratislava: Obzor, 1981. p. 9–36.

JESENSKÝ, Vít. 2015. K teorii památkové hodnoty a hodnocení památek. In *Památky západních Čech V-2015*, Plžeň: Národní památkový ústav, územní odborné pracoviště v Plzni, 2015, p. 56–64. ISBN 978-80-85035-46-9.

JESENSKÝ, Vít. 2019. Kontexty památkové péče jako kategorie utváření její teorie. In HORÁČEK, Martin. et al. 2019. *Teoretické základy památkové péče na prahu 21. století*. Praha: Artefactum, nakladatelství Ústavu dějin umění AV ČR, v.v.i. p. 47–64. ISBN 978-80-88283-27-0.

JOHNSTON, Kevin et al. 2003. *ArcGIS 9. Using ArcGIS® Geostatistical Analyst* [online]. New York: Copyright © 2003 ESRI, 2003. s. 306 [Accessed on: 18.7.2024]. Available at: https://www.researchgate.net/publication/200043204_Using_ArcGIS_geostatistical_analyst

JONES M. David. 2011. *Environmental Archaeology. A Guide to the Theory and Practice od Methods, from Sampling and Recovery to Post-excavation (second edition)* [online]. English Heritage [Accessed on: 5.3.2024] Available at: https://www.researchgate.net/publication/295912767_Environmental_Archaeology_A_guide_to_the_theory_and_practice_of_methods_from_sampling_and_recovery_to_post-excavation

JORDAN-PALOMAR, Isabel / TZORTZOPOULOS, Patricia / GARCÍA-VALLDECABRES, Jorge / PELLICER, Eugenio. 2018. *Protocol to manage heritage-building interventions using heritage building information modelling* (HBIM) [online]. *Sustainability*, 10(3), 2018 [Accessed on: 13.3.2024]. Available at: https://www.mdpi.com/2071-1050/10/4/908

JURKOVIČ, M. 1969. Fyloxéra viničová. In *Agrikultúra*. vol. 8, p. 115–129.

KANTNER, John. 2000. Realism vs. Reality: Creating Virtual Reconstructions of Prehistoric Architecture [online]. In *Virtual Reality in Archaeology*, eds. . BARCELO, J - FORTE, A. M. / SANDERS, D. H., Oxford: Archaeopress, 2000. p. 47–52 [Accessed on: 18.7.2024]. Available at: https://www.researchgate.net/publication/240305149_Realism_vs_reality_Creating_virtual_reconstructions_of_prehistoric_architecture

KARLSSON, Christina. 2013. *Visualizing Archeology.* [online]. Lund: Lund University [Accessed on: 2018-04-07]. Available at: http://lup.lub.lu.se/luur/download?func=downloadFile&recordOId=3800424&fileOId=3800425<

KAZIMÍR, Štefan. 1986. *Pestovanie viniča a produkcia vína na Slovensku*. Bratislava: VEDA, vydavateľstvo Slovenskej akadémie vied, 1986. 327 p.

KENSEK, Karen / DODD, Lynn Swartz / CIPOLLA, Nicholas. 2004. Fantastic reconstructions of the fantastic? Tracking and presenting ambiguity, alternatives, and documentation in virtual worlds [online]. In *Automation in Construction*, vol. 13, 2004 p. 175–186 [Accessed on: 5.7.2024]. Available at: DOI: 10.1016/j.autcon.2003.09.010

KLIR, George / YUAN, Bo. 1995. *Fuzzy Sets and Fuzzy Logic: Theory and Applications.* Upper. Saddle River: Prentice Hall PTR. 591 p. ISBN 0-13-101171-5.

KOLEJKA, Jaromír. 2013. *Nauka o krajině pro studujícií geografie magisterských učitelských odborů*. Brno: Pedagogická fakulta Masarykovy univerzity v Brně, 2013. 123 p. ISBN 978-80-904785-1-0.

KORČÁK, Pavel. 1990. Diskuse o teorii a metodologii památkové péče. In *Památky a příroda*, 1990, vol. 50, no. 1, p. 65–74.

KOČICKÝ, D. / IVANIČ, B. 2011. *Geomorfologické členenie Slovenska*. [online]. Bratislava: ŠGDÚŠ, 2011. [Accessed on: 13.8.2024]. Available at: https://apl.geology.sk/mapportal/ img/pdf/tm19a.pdf.

KUHN, Thomas S. 1970. *The Structure of Scientific Revolutions*. Chicago: The University of Chicago, 1970. 222 p. ISBN 81-80-79-78.

KUNA, Martin et al. 2004. *Nedestruktivní archeologie*. Praha: Academia, 2004. 555 p. ISBN 80-200-1216-8.

KUNA, Martin. 1994. Archeologický průzkum povrchovými sběry. In *Zprávy české archeologické společnosti, supplement 23*, 1994. Praha: Sdružení archeologů Čech, Moravy a Slezka. 111 p.

KUTLÍK, František. 1981. Príspevok k dejinám pamiatkovej legislatívy na Slovensku. In *Jubilejný zborník: 30 rokov Slovenského ústavu pamiatkovej starostlivosti a ochrany prírody v Bratislave*. Bratislava: Obzor, 1981. p. 37–64.

KUČERA, Zdeněk / SEMOTANOVÁ, Eva. 2018. Historická krajina jako "theatrum mundi". In SEMOTANOVÁ, Eva / CHROMÝ, Pavel / KUČERA, Zdeněk. *Historická geografie: tradice a modernita*. Praha, 2018, p. 141–178.

KUČOVÁ, Věra / DOSTÁLEK, Jiří / EHRLICH, Marek / KUČA, Karel / PACÁKOVÁ, Božena. 2014. *Metodika tvorby standardizovaného záznamu krajinné památkové zóny*. Praha: Národní památkový ústav, 2014. 136 p. ISBN 978-80-7480-0221-1.

KÜÇÜKDEMIRCI, Melda / SARRIS, Apostolos. 2022. GPR Data Processing and Interpretation Based on Artificial Intelligence Approaches: Future Perspectives for Archaeological Prospection [online]. In *Remote Sens*, vol. 14, 2022. p. 1–11 [Accessed on: 18.7.2024]. Available at: https://www.mdpi.com/2072-4292/14/14/3377

LAMBA, Manika / MADHUSUDHAN, Margam. 2022. Predictive Modeling [online]. In *Text Mining for Information Professionals: An Uncharted Territory*, Springer Nature, 2022. p. 213–242. ISBN 978-3-030-85084-5 [Accessed on: 17.7.2024]. Available at: https://www.researchgate.net/publication/360113873_Predictive_Modeling

LAUMAIN, Xavier / ESTEVE MUÑOZ, Vicente Francisco / LÓPEZ SABATER, Ángela. 2023. Studies of the Nolla Palace. From survey to HBIM Managment. In LÓPEZ GONZÁLEZ, Concepción. 2023. *EGE Revista de Expresion Gráfica en la Edificación*, vol. 19, 2023. p. 96–116. ISSN 2605-082X.

LENDA, Grzegorz / UZNAŃSKI, Andrzej / STRACH, Michał –LEWIŃSKA, Paulina. 2016. Laser scanning in Engineering Surveying: Methods of measurement and modeling of structures [online]. In *Reports on Geodesy and Geoinformatics*, vol. 100/2016, p. 109–130 [Accessed on: 5.3.2024]. Available at: https://intapi.sciendo.com/pdf/10.1515/rgg-2016-0010

LIESKOVSKÝ, Juraj / LIESKOVSKÝ, Tibor / HLADÍKOVÁ, Katarína/ ŠTEFUNKOVÁ, Dagmar. 2022. Potential of airborne LiDAR data in detecting cultural landscape features in Slovakia. In *Landscape Research*, vol. 47, no. 5, 2022. p. 539–558.

LILLESAND, Thomas / KIEFER, Ralph / CHIPMAN, Jonathan. 2014. *Remote Sensing and Image Interpretation, 7th Edition*. WILLEY, 2014. 736 p. ISBN 978-1-119-12846-5.

LIPSKÝ, Zdeněk. 1999. *Krajinná ekologie pro studenty geografických odborů*. Praha: Karolinum/nakladatelství Univerzity Karlovy, 1999. 129 p. ISBN 382-181-97.

LIPSKÝ, Zdeněk. 2000. *Sledování změn v kulturní krajině. Učební text pro cvičení z předmětu Krajinná ekologie*. Kostelec nad Černými lesy: ÚAE LF ČZU, 2000. 71 p. ISBN 80-213-0643-2.

LIU, Junshan / WILLKENS, Danielle, S. / FOREMAN, Graham. 2022. An introduction to technological tools and process of heritage Building information modeling (HBIM). In *EGE Revista de Expresión Gráfica en la Edificación*, no. 16, Valencia: Universitat Politécnica de València, 2022. p. 50–65. ISSN: 2605-082X.

London Charter for the computer-based visualization of cultural heritage [online]. © 2009 LONDONCHARTER.ORG [Accessed on: 22.7.2024]. Available at: https://londoncharter.org

LOPEZ-MENCHERO, Victor Manuel / GRANDE, Alfredo. 2011. *The Principles of the Seville Charter* [online]. CIPA Heritage Documention, 2011 [Accessed on: 25.7.2024]. Available at: https://www.cipaheritagedocumentation.org/wp-content/uploads/2018/12/López-Menchero-Grande-The-principles-of-the-Seville-Charter.pdf

LOWENTHAL, David. 2005. Natural and Cultural Heritage. In *International Journal of Heritage Studies*, 2005, vol. 11, no. 1. p. 81–92. ISSN 1352-7258.

MACEACHREN, Alan et al. 2005. Visualizing Geospatial Information Uncertainty: What We Know and What We Need to Know. In *Cartography and Geographic Information Science*, vol. 32, no. 3. p. 139–160. ISSN 1523-0406.

Mapový klient ZBGIS: Verzia 5.11.2 [online]. Úrad geografie a kartografie Slovenska, 2018 [Accessed on: 30. 7. 2024]. Available at: https://zbgis.skgeodesy.sk/mkzbgis/sk/zakladna-mapa?pos=48.232761,19.422368,14

Maps Arcanum: historical maps of several countries [online]. Arcanum adatabázis kft. [Accessed on: 30. 7. 2024]. Available at: https://maps.arca- num.com/en/

MARA, Hubert. 2022. Digital Archaeology [online]. In *It - Information Technology*, vol. 64, no. 6, 2022. p. 193–194. [Accessed on: 17.7.2024]. Available at: https://www.researchgate.net/publication/365079051_Digital_Archaeology/citations.

MARČIŠ, Marián. 2019. *Automatizované fotogrametrické metódy v procese digitalizácie kultúrneho dedičstva*. STU, Stavebná fakulta, Bratislava 2019. 113 p. ISBN 978-80-227-4895-7.

MICHALOVÁ, Jaroslava / MICHAL, Pavel. 1980. *Geografia okresu Veľký Krtíš*. Veľký Krtíš: Vydavateľstvo Osveta, 1980. 288 p.

MICHALÍK, Tomáš. 2013. *Ako sa dohodnúť s pamiatkarmi?* [online]. ASB, p. 11–12, 2013 [Accessed on: 5.2.2024]. Available at: https://www.heritageconsulting.sk/wp-content/uploads/AZA_Michalik.pdf

MICHELE, Melchiorri et al. 2019. *Multitemporal Grid Based Analysis of the Global Human Settlement Layers*. Luxembourg: Publications Office of the European Union. 29 p. ISBN 978-92-79-98754-0.

MIKHAIL M., Edward / BETHEL S., James / McGLONE, Chris. 2001. *Začiatok formulára Spodná časť formulára*

MIKLÓS, László / IZAKOVIČOVÁ, Zita. 1997. *Krajina ako geosystém*. Bratislava: Vydavateľstvo VEDA, 1997. 152 p. ISBN 80-224-0519-1.

MITCHELL, Andy / GRIFFIN, Lauren Scott. 2021. *The Esri Guide to GIS Analysis, Volume 2: Spatial Measurements and Statistics, second edition*. Redlands, Kalifornia: ESRI, 2021. 288 p. ISBN 978-15894-860-89.

MITTELSTADT, Brent. 2019. Principles alone cannot guarantee ethical AI [online]. In *Nature Machine Intelligence*, vol. 1, no. 11, 2019. p. 501–507

[Accessed on: 18.7.2024]. Available at: https://www.nature.com/articles/s42256-019-0114-4#citeas

MLINKAUSKIENĖ, Aušra / JANKAUSKAITĖ-JUREVIČIENĖ, Laura. 2019. Investigation of Heritage Building Information Modeling Possibilities in Lithuania. In *Architecture and Urban Planning*, [online]. 2019, vol. 15, [Accessed on: 10.3.2024]. Available at: doi:10.2478/aup-2019-0007

MÅRTENSSON, Ulrik. 2011. *Introduction to Remote Sensing and Geographical Information Systems* [online]. Lund: Lund University, 2011. [Accessed on: 5.3.2024]. Available at: https://www.nateko.lu.se/sites/nateko.lu.se.sv/files/remote_sensing_and_gis_20111212.pdf

NARIADENIE EURÓPSKEHO PARLAMENTU A RADY (EÚ) č. 1308/2013 zo 17. decembra 2013. [online]. [Accessed on: 10.11.2021]. Available at: https://eur-lex.europa.eu/legal-content/SK/TXT/?uri=CELEX:32013R1308

NEUSTUPNÝ, Evžen. 2007. *Metoda archeologie*. Plzeň: Vydavatelství a nakladatelství Aleš Čeněk, s.r.o., 2007. 206 p. ISBN 978-80-7380-075-8.

NIJHUIS, Steffen et al. 2011. Exploring the Visual Landscape. Advancs in Physiognomic Landscape Research in the Netherlands. In *Research in Urbanism series*, vol. 2, Delf University of Technology, Amsterdam, 2011. 336 p. ISSN 1875-0192.

NIJHUIS, Steffen. 2016. *Applications of GIS in landscape design research* [online]. Research. In *Urbanism Series*, vol. 4, no. 1, p. 43–56, 2016 [Accessed on: 10.3.2024]. Available at: https://rius.ac/index.php/rius/article/view/110/82

NOUR EL DIN, Mohamed / GRUSSENMEYER, Pierre / KOEHL, Mathieu. 2000. Modelling historic sites and monuments in 3D heritage information systems by combining aerial and architectural photogrammetry [online]. In *XIXth Congress of the International Society for Pho- togrammetry and Remote Sensing* (ISPRS), Geoinformation for all, Amsterdam, 2000, Netherlands. p. 1–8 [Accessed on: 18.7.2024]. Available at: https://www.academia.edu/23146550/Modelling_historic_sites_and_monuments_in_3D_heritage_information_systems_by_combining_aerial_and_architectural_photogrammetry

NÉMETHOVÁ, Jana / KRAJČI, Filip. 2021. Vinohradníctvo a vinárstvo Slovenska v komparácii so štátmi Európskej únie po roku 2004. In *Geografické informácie*, vol. 25, no. 2, 2021. p. 34–49. ISSN 1337-9453

OLIVITO, Ricardo / TACCOLA, Emanuele. 2014. 3D Modelling in the Agora of Segesta: Techniques and data interpretation [online]. In *Archeologia e Calcolatori*, vol. 25, 2014. p. 175–188 [Accessed on: 18.7.2024]. Available at: https://www.archcalc.cnr.it/indice/PDF25/09_Olivito_Taccola.pdf

ORENI, Daniela et al. 2014. Survey turned into HBIM: the Restoration and the work involved concerning the Basilica di Collemaggio after the Earthquake (L'aquila) [online]. In *ISPRS Annals of the Photogrammetry, Remote Sensing and Spatial Information Sciences*, ISPRS Technical Commission V Symposium, 23–25 June 2014, Riva del Garda, Italy, vol. II, no. 5, 2014 [Accessed on: 18.3.2024]. Available at: https://re.public.polimi.it/bitstream/11311/885957/3/Fabrizio%20Banfi%20-%202014_SURVEY%20TURNED%20INTO%20HBIM.pdf

OROSOVÁ, Martina. 2003. Dielo manželov Menclovcov v oblasti záchrany kultúrneho dedičstva Slovenska. In *Zborník Slovenského národného múzea: Etnografia XCVII*. Martin: Slovenské národné múzeum/Etnografické múzeum, 2003. p. 60–74. ISBN 80-8060-113-5.

OTT, Thomas / SWIACZNY, Frank. 1998. The analysis of cultural landscape change: a GIS approach for handling spatio-temporal data. In *History and Computing*, vol. 10, no. 1–3, 1998. International Journal of Humanities and Arts Computing. p. 37–49. ISSN 0957-0144.

OŤAHEĽ, Ján. 1996. Krajina/pojem a vnem. In *Geografický časopis*, 1996, vol. 48, no. 3–4, p. 241–253. ISSN 2453-8787.

PIETRONI, Eva / FERDANI, Daniele. 2021. Virtual Restoration and Virtual Reconstruction in Cultural Heritage: Terminology, Methodologies, Visual Representation Techniques and Cognitive Models [online]. In *Information*, 2021, vol. 12, no. 167, 2021 [Accessed on: 25.3.2024]. Available at: https://www.mdpi.com/2078-2489/12/4/167

PITOŇÁK, Martin. 2017. *Využitie BIM (ang. Building Information Modeling) v cestnom hospodárstve na Slovensku* [online]. Žilina: CEDS, Stavebná fakulta, Žilinská univerzita, 2017, 67 p. [c Accessed on: 13.3.2024]. Available at: https://www.ssc.sk/files/documents/technicke-predpisy/rozborove_ulohy/2017_ru_bim_ch.pdf

PLAČEK, Miroslav / BÓNA, Martin. 2007. *Encyklopédia slovenských hradov*. Bratislava: SLOVART, spol. s. r. o. 391 p. ISBN 978-90-8085-287-0.

POLLARD, Mark / BATT, Cathy / STERN, Ben / M. M. YOUNG, Suzzane. 2006 *Analytical Chemistry in Archaeology*. Cambridge: University Press, 2006. 419 p. ISBN 978-0-511-349942.

PUYOL-GRUART, Josep. 2002. *Computer Science, Artificial Intelligence and Archaeology* [online]. [Accessed on: 18.7.2024]. Available at: https://www.researchgate.net/publication/2834381_Computer_Science_Artificial_Intelligence_and_Archaeology

QUATTRINI, Ramona / PIERDICCA, Roberto / MORBIDONI, Christian. 2017. Knowledge-based data enrichment for HBIM: Exploring high-quality models using the semantic-web. In *Journal of Cultural Heritage*, vol. 28, 2017, s. 129–139. ISSN 1296-2074.

Rada Európy *European Treaty Series* - No. 176 [online]. Council of Europe, Florence 2000 [Accessed on: 6.2.2024]. Available at: https://www.minzp.sk/files/postupy-a-ziadosti/ochrana-prirody/medzinarodne-dohovory/o-krajine/4-edokslovverzia.pdf

Register NNKP. Register nehnuteľných národných kultúrnych pamiatok [online]. Pamiatkový úrad Slovenskej republiky [Accessed on: 17.7.2024]. Available at: https://www.pamiatky.sk/evidencie-a-registre/register-nnkp

REILLY, Paul 1990. *Towards a virtual archaeology* [online]. IBM UK Scientific Centre, Athelstan House, St. Clement Street, Winchester, S023 9DR, p. 133–139 [Accessed on: 17.7.2024]. Available at: https://www.researchgate.net/publication/235902686_Towards_a_virtual_archaeology

REILLY, Paul. 1992. Three-dimensional modelling and primary archaeological data [online]. In REILLY P., RAHTZ S. (eds.), *Archaeology and the Information Age. A Global Perspective*, London-New York, Routledge, 1992, p. 147–173 [Accessed on: 18.7.2024]. Available at: https://archaeologicalcomputing.lincei.it/reilly-p-three-dimensional-modelling-and-primary-archaeolog

RENFREW, Colin / BAHN, Paul. 2000. Archaeology. Theories Methods and Practice. London: Thames & Hudson, 2000. 672 p. ISBN 0-500-28147-5.

RODRÍGUEZ-GONZÁLVEZ, Pablo et al. 2017. 4D Reconstruction and Visualization of Cultural Heritage: Analyzing our Legacy through time [online]. In *The International Archives of the Photogrammetry, Remote Sensing and Spatial Information Sciences*, vol. XLII–2, no. 3, 2017 3D Virtual Reconstruction and Visualization of Complex Architectures, March 2017, Nafplio, Greece, p. 1–3 [Accessed on: 8.4.2024]. Available at: https://isprs-archives.copernicus.org/articles/XLII-2-W3/609/2017/isprs-archives-XLII-2-W3-609-2017.html

RODRÍGUEZ-GONZÁLVEZ, Pablo et al. 2018. 4D Reconstruction of Cultural Heritage Sites [online]. In REMONDINO, F. - GEORGOPOULOS, A. - GONZÁLEZ-AGUILERA, D.; AGRAFIOTIS, P.; Eds., *Latest Developments in Reality-Based 3D Surveying and Modelling*; MDPI: Basel, Switzerland, 2018; p. 119–140 [Accessed on: 8.4.2024]. Available at: https://d1wqtxts1xzle7.cloudfront.net/81600424/4D_Reconstruction_of_Cultural_Heritage_Sites-libre.pdf?1646257869=&response-content-disposition=inline%3B+filename%3D4D_Reconstruction_of_Cultural_Heritage_S.pdf&Expires=1712592451&Signature=doAm8dZ7PCiAxfJFVGlMUCD26ZHz~HtNU8pObFMBsXS9j0OPrCmj48tyqXZ9UzbP9EUiWVBwxMzRCYpgx6vd0Mn9DQT1KSwRL7sY5k08kg35stnyn~B4luOISG51q4iDmck1ymQSD3RcleA3mvZWhQ~XTaYpIMegDGxzFb9Gf7ceHex5n7vRO~9FYbQD-3Q6-zjyNDvaX5U5z7fwDFcLKytThstGtkBOrhXBmarjQ9iXsQMeETtEORA3-4iUoSwxV-deAoOvSGCYGqBU2UsDtwoof5OOLj~TtOK7SmNo5IIKAVdZgDuecc61fNTP-FgZMof9q439719BpkGdiWoGOw__&Key-Pair-Id=APKAJLOHF5GGSLRBV4ZA

RODRÍGUEZ-MORENO, Concepcion et al. 2016. From point cloud to BIM: an integrated workflow for documentation, research and modelling of architectural heritage [online]. In *Survey Review*, 2016. p. 1–20 [Accessed on: 18.3.2024]. Available at: https://www.researchgate.net/publication/311355443_From_point_cloud_to_BIM_an_integrated_workflow_for_documentation_research_and_modelling_of_architectural_heritage

ROMELE, Alberto / SEVERO, Marta / FURIA, Paolo. 2018. Digital hermeneutics: from interpreting with machines to interpretational machines [online]. In *AI & SOCIETY*, vol. 35, no. 3, 2018 p. 73–87 [Accessed on: 17.7.2024]. Available at: https://www.researchgate.net/publication/326091804_Digital_Hermeneutics_From_Interpreting_with_Machines_to_Interpretational_Machines

ROUSSOU, Maria / DRETTAKIS, George. 2003. Photorealism and non-photorealism in virtual heritage representation [online]. In eds. CHALMERS, A / ARNOLD, D. / NICCOLUCCI, F. *Proceedings of VAST International Symposium on Virtual Reality, Archaeology and Cultural Heritage including the First References 201 Eurographics Workshop VAST '03*, 2003,

p. 51–60 [Accessed on: 3.8.2024]. Available at: http://www-sop.inria.fr/reves/Basilic/2003/RD03b/RoussouDrettakis03.pdf

RZASA, Krysztof / OGRYZEK, Marek / CISKI, Mateusz. 2019. Application of GIS Tools in Spatial Distribution Modeling of Historical Monuments [online]. In *Geomatics and Environmental Engineering*, vol. 13, no. 2, 2019 [Accessed on: 17.7.2024]. Available at: https://www.gaee.agh.edu.pl/index.php/gaee/article/view/151

RÁBIK, Vladimír / LABANC, Peter / TIBENSKÝ, Martin. 2013. *Historická geografia*. Trnava: Filozofická fakulta Trnavskej univerzity v Trnave, 2013. 43 p. ISBN 978-80-8082-643-7.

SADREHAGHIGHI, Ideen. *Computer Aided Design (CAD)* [online]. Anatomy of commercial CAD Systems, Annapolis, MD, 2022 [Accessed on: 17.7.2024]. Available at: https://www.researchgate.net/publication/358140456_Computer_Aided_Design_CAD

SALAŠOVÁ, Alena / SEDLÁČEK, Jozef / TRPÁKOVÁ, Lenka / MARTINKOVÁ KUCHYŇKOVÁ Hana. 2015. *Metodika monitoringu stavu krajinné památkové zóny*. Brno: Mendelova univerzita v Brně 2015. 88 p. ISBN 978-80-7509-359-2.

SALVIA, Stefano. 2007. Reconstruction of Scientific Instruments and Experimental History of Science [online]. In VACKIMES, S. / WELTERSBACH, K. (eds.). *Wandering Seminar on Scientific Objects*, MPIWG, Berlin 2007, p. 81–108 [Accessed on: 25.3.2024]. Available at: https://www.academia.edu/4398321/Reconstruction_of_Scientific_Instruments_and_Experimental_History_of_Science

SÁNCHEZ, Gonzalo Martín. 2024. *Development of a 3D model file for preservation of monuments* [online]. [Accessed on: 18.7.2024]. Available at: https://www.academia.edu/12363819/Development_of_a_3D_model_file_for_preservation_of_monuments

SCHULZE-HORSEL, Michael. 2007. 3D landmarks: generation, characteristics and applications [online]. In REMONDINO, Fabio - EL-HAKIM, Sabry, eds. 2007. *3D-ARCH 2007*: “Virtual Reconstruction and Visualization of Complex Architectures” 12–13 July 2007, ETH Zurich, Switzerland [Accessed on: 5.2.2024]. Available at: internete: https://www.isprs.org/proceedings/xxxvi/5-w47/pdf/schulze-horsel.pdf

SCHÄFER, Una Ulrike. 2018. A Brief Introduction [online]. In *International Journal for Digital Art History*, no. 3, 2018. p. 87–105 [Accessed on: 18.7.2024]. Available at: https://journals.ub.uni-heidelberg.de/index.php/dah/article/view/32703/43446

SEDLÁČEK, Jozef / SALAŠOVÁ, Alena / TRPÁKOVÁ, Lenka. 2015. *Metodika hodnocení vlivů na krajinné památkové zóny*. Brno: Mendelova univerzita v Brně 2015. 72 p. ISBN 978-80-7509-360-8.

SEMOTANOVÁ, Eva. 1998. *Historická geografie českých zemí*. Praha: Historický ústav AV ČR, 1998. 293 p. ISBN 80-85268-73-6.

SIFNIOTIS, Maria et al. 2007. Influencing Factors on the Visualisation of Archaeological Uncertainty [online]. In ARNOLD, D. - NICCOLUCCI, F. and CHALMERS, A. eds. *The 8th International Symposium on Virtual Reality and Intelligent Cultural Heritage*, Brighton, November 26–30, 1–4. Aire-La-Ville: Eurographics Association, 2007 [Accessed on: 7.7.2024]. Available at: https://www.researchgate.net/publication/220955336_Influencing_Factors_on_the_Visualisation_of_Archaeological_Uncertainty

SINGH, Ripudaman. 2021. *Re-envisioning Remote Sensing. Applications Perspectives from Developing Countries*. Boca Raton: CRC Press, 2021. 347 p. ISBN 978-0-367-50239-3.

SLOBODOVÁ-NOVÁKOVÁ, Katarína. 2020. *Vinohradníctvo a vinárstvo na Slovensku. Vybrané kapitoly z etnografie materiálnej kultúry*. Trnava: Univerzita sv. Cyrila a Metoda v Trnave, 2020. 91 p. ISBN 978-80-572-0059-8.

STEHLÍK, Miloš. 2018. *Kam kráčíš, památková péče?* Brno: Masarykova univerzita 2018. 188 p. ISBN 978-80-7485-158-2.

STROTHOTTE, Thomas / MASUCH, Maic / ISENBERG, Tobias. 1999. Visualizing Knowledge about Virtual Reconstructions of Ancient Architecture [online]. In LEWIS, T. ed. *Proceedings Computer Graphics International. The Computer Graphics Society*, IEEE Computer Society, 7–11 June, 1999, Los Alamitos,1999, p. 36–43 [Accessed on: 2.7.2024]. Available at: https://www.researchgate.net/publication/3806256_Visualizing_Knowledge_about_Virtual_Reconstructions_of_Ancient_Architecture

SVOBODOVÁ, Helena. 1986. Použití metody minerální separace pylových zrn a spór na archeologické kulturní vrstvy jižní Moravy. In *Archeologické rozhledy 38*, Praha: Nakladatelství ČSAV, 1986. p. 283–287.

The ICOMOS Landscape Working Group Heritage [online]. Heidelberg University [Accessed on: 6.2.2024]. Available at: https://www.google.com/url?sa=t&rct=j&q=&esrc=s&source=web&cd=&ved=2ahUKEwjHuce_k5eEAxUz9wIHHUQAC7AQFnoECBQQAQ&url=https%3A%2F%2Fjournals.ub.uni-heidelberg.de%2Findex.php%2Ficomoshefte%2Farticle%2Fview%2F22585%2F16346&usg=AOvVaw2HcxT2B_vthmSCxp-1rbN6&opi=89978449

The Operational Guidelines for the Implementation of the World Heritage Convention [online]. UNESCO World Heritage Covention [Accessed on: 6.2.2024]. Available at: https://whc.unesco.org/en/guidelines/

THEMISTOCLEOUS, Kyriaos. 2019. The Use of UAVs for Cultural Heritage and Archaeology. In DIOFANTOS, G. *Hadjimitsis. Remote Sensing for Archaeology and Cultural Landscapes: Best Practices and Perspectives Across Europe and the Middle East.* Springer Nature, 2019. 308 p. ISBN 978 3030-10978-3.

TOMEČEK, Oto. 2007. Poznámky k vývoju a aktuálnemu stavu historicko-geografického výskumu na Slovensku. In *Historická geografie,* vol. 34. Praha: Historický ústav AV ČR, 2007, p. 21–43. ISSN 0323-0988.

TOMOVIČOVÁ, Jana / REICHARDTOVÁ, Juliana. 2023. Archívne a knižničné zdroje na výskum mlynov v Archíve Pamiatkového úradu SR. In BARTOŠÍKOVÁ, Tereza et al. *K dejinám mlynov/výskum a mapovanie.* Pamiatkový úrad Slovenskej republiky, Vodné mlyny Oblazy, n.o. Zborník príspevkov zo seminára Zborník príspevkov zo seminára 14.12.2021, 2023. p. 177–128. ISBN 978-80-89175-97-0.

TUFTE, Edward. R. 2001. *The Visual Display of Quantitative Information [online].* Cheshire: Graphics Press, 2001 [Accessed on: 18.7.2024]. Available at: http://faculty.salisbury.edu/~jtanderson/teaching/cosc311/fa21/files/tufte.pdf

UETNOFOLK, Digitálne záznamy. [online]. Bratislava: Ústav etnológie a sociálnej antropológie SAV. [Accessed on: 6.7.2024]. Available at: https://uetetnofolk.eu/index.php/main/records

UHLÍK, Jan. 2019. Hodnota celistvosti v kontextu hodnot památek, její formy a význam. In. HORÁČEK, Martin. et al. 2019. *Teoretické základy památkové péče na prahu 21. století.* Praha: Artefactum, nakladatelství Ústavu dějin umění AV ČR, v.v.i. p. 117–150. ISBN 978-80-88283-27-0.

UNGER, Jiří. 2019. *Možnosti využití 3D rekonstrukčních počítačových vizualizací pro archeologii*. Dizertačné práca: Univerzita Karlova, Praha 2019. 249 p.

URBAN, Peter. 2010. *Zlatá kniha Hontu. Turany*: Matica slovenská, 2010. 195 p. ISBN 978-80-7090-955-3.

VALERO, Enrique, et al. 2017. Evaluation of historic masonry: towards greater objectivity and efficiency. In YUSUF, Arayici et al. *Heritage Building Information Modeling*. London & New York: Taylor & Francis, 2017. p. 75–101. ISBN 978-1-138-64568-4.

VASS, Dionýz / KONEČNÝ, Vlastimil/ŠEFARA, Jan et al. 1979. *Geologická stavba Ipeľskej kotliny a Krupinskej planiny*. Bratislava: Geologický ústav Dionýza Štúra, 1979. 277 p.

VAUGHN, Danny M. 2019. *Geostatistical Analyst: An ArcGIS Extension* [online]. 2019 [Accessed on: 17.7.2024]. Available at: https://www.academia.edu/39168245/GEOSTATISTICAL_ANALYST

VENCL, Slavomir. 1995. K otázce věrohodnosti svědectví povrchových průzkumů. In *Archeologické rozhledy*, no. 47, Praha: Nakladatelství ČSAV, 1995. p. 11–57.

VESELÝ, Jan. 2014. *Měřická dokumentace historických staveb pro průzkum v památkové péči*. Praha: Národní památkový ústav, 2014. 127 p. ISBN 978-80-86516-79-0.

VILLEGAS-GALAVIZ, Carolina / MARTIN, Kirsten. 2023. Moral distance, AI, and the ethics of care [online]. In *AI & SOCIETY*, 2023 [Accessed on: 18.7.2024]. Available at: https://link.springer.com/article/10.1007/s00146-023-01642-z#citeas

VĚŽNÍK, Antonín / KYSELÝ, Jakub. 2019. Geografická analýza produkce révy vinné na území Jihomoravského kraje. In *Geografické informácie*. vol. 23, no. 2, 2019, p. 66–82. ISSN 1337-9453.

WENG, Qihao. 2009. *Remote Sensing and GIS Integration: Theories, Methods, and Applications: Theory, Methods, and Applications*. Terre Haute: McGraw Hill, 2009. 416 p. ISBN 978-0071606530.

WITTUR, Joyce. 2013. *Computer-Generated 3D-Visualisations in Archaeology: Between added value and deception*. (BAR International Series 2463). Oxford: Archeopress. 297 p. ISBN 978-14073-107-1-8.

World Heritage List [online]. The city of Luxembourg: its old quarters and fortifications, Grand Duchy of Luxembourg, Luxembourg 1993 [Accessed on: 6.2.2024]. Available at: https://www.google.com/url?sa=t&rct=j&q=&esrc=s&source=web&cd=&ved=2ahUKEwj595OklJeEAxWP3AIHHXHtDXYQFnoECBkQAQ&url=https%3A%2F%2Fwhc.unesco.org%2Fdocument%2F154048&usg=AOvVaw1K55VEHmZ50hOtHb5GTcwY&opi=89978449

YANG, Chen / HAN, Feng / WU, Hangbin / CHEN, Zhuo. 2019. Heritage Landscape Information Model (HLIM): towards a contextualised framework for Digital Landscape Conservation in China [online]. In *Arch. Photogramm. Remote Sens.* Spatial Inf. Sci., vol. XLII–2 / W15, p. 1221–1227 [Accessed on: 18.3.2024]. Available at: https://isprs-archives.copernicus.org/articles/XLII–2-W15/1221/2019/

ZHANG, Hui et al. 2014. The Application of GIS 3D Modeling and Analysis Technology in Real Estate Mass Appraisal - Taking landscape and sunlight factors as the example [online]. In *The International Archives of the Photogrammetry, Remote Sensing and Spatial Information Sciences*, vol. 40, no. 4, 2014. p. 363–367 [Accessed on: 18.7.2024]. Available at: https://www.researchgate.net/publication/263018887_The_Application_of_GIS_3D_Modeling_and_Analysis_Technology_in_Real_Estate_Mass_Appraisal_-_Taking_landscape_and_sunlight_factors_as_the_example

ZUK, Torre. 2008. *Visualizing Uncertainty* [online]. PhD. Thesis, Department of computer science, The University of Calgary, Calgary, Alberta, 2008, 311 p. [Accessed on: 18.7.2024]. Available at: https://www.researchgate.net/publication/286242934_Visualizing_Uncertainty

ZUK, Torre / CARPENDALE, Sheelagh. 2006. *Theoretical Analysis of Uncertainty visualizations* [online]. Proceedings of SPIE 6060, Visualization and Data Analysis 2006. 15–19. January 2006, San Jose, California, USA, 2006. p. 1–14 [Accessed on: 20.7.2024]. Available at: https://www.researchgate.net/publication/228730951_Theoretical_analysis_of_uncertainty_visualizations

ČECH, Vladimnír / KROKUSOVÁ, Juliana. 2013. *Antropogénna geomorfológia (antropogénne formy reliéfu)*. Prešov: Katedra geografie a regionálneho rozvoja, 2013. 179 p. ISBN 978-80-555-1037-8.

ŠANTRŮČKOVÁ, Markéta. 2019. Identifikace hodnot historické kulturní krajiny a jejich ochrana pomocí krajinných památkových zón. In HORÁČEK, Martin. et al. 2019. *Teoretické základy památkové péče na prahu*

21. století. Praha: Artefactum, nakladatelství Ústavu dějin umění AV ČR, v.v.i. p. 188. ISBN 978-80-88283-27-0.

ŠPULEROVÁ, Jana/ŠTEFUNKOVÁ, Dagmar / DOVODSKÁ, Marta et al. 2017. *Historické štruktúry poľnohospodárskej krajiny Slovenska*. Bratislava: VEDA Vydavateľstvo SAV, 2017. 144 p. ISBN 978-80-224-1570-5.

ŠTEFUNKOVÁ, Dagmar / KRNÁČOVÁ, Zdena. 2020. Odkaz kultúrneho dedičstva v krajine v podobe historických agrárnych štruktúr vinohradníckej krajiny. In *Acta Environmntalica Universitatis Comenianae* (Bratislava), 2012, vol. 20, no. 1, p. 106–109. ISSN 1335-0285.

ŠTULC, Josef. 2010. *Úvod do studia památkové péče* [online]. Praha: EU: Investujeme do vaší budoucnosti Evropský sociální fond, 2010 [Accessed on: 5.2.2024] Available at: http://duoppa.ff.cuni.cz/shared/files/Pamatkova%20pece/doc.%20Stulc%20Vyvoj%20PP%2009%2011%202010.pdf

ŽIGRAI, Florin. 1999. Význam časopriestoru pri transformácii kultúrnej krajiny (Vybrané teoreticko-metodologické aspekty). In NOVODOMEC, Rudolf et al. *Geografické štúdie Nr. 6: Stredné Slovensko/transformácia prírodnej a kultúrnej krajiny*. Banská Bystrica, 1999, p. 51–60. ISBN 80-8055-471-4.

ŽIGRAI, Florin. 2000. Dimenzie a znaky kultúrnej krajiny. In *Životné prostredie*, 2000, vol. 34, no. 5, p. 229–233. ISSN 0044-4863.

www.ingramcontent.com/pod-product-compliance
Ingram Content Group UK Ltd.
Pitfield, Milton Keynes, MK11 3LW, UK
UKHW021829100926
8556IPUK00004B/239

9 783631 936221